Irmfried Hartmann
Werner Lange
Rainer Poltmann

Robust and Intensitive Design of Multivariable Feedback Systems

Advances in Control Systems and Signal Processing

Editor: Irmfried Hartmann

Volume 6

Volume 1: Erhard Bühler and Dieter Franke
Topics in Identification and Distributed Parameter Systems

Volume 2: Hubert Hahn
Higher Order Root-Locus Technique with Applications in Control System Design

Bernhard Herz
A Contribution about Controllability

Volume 3: Günter Ludyk
Time-Variant Discrete-Time Systems

Volume 4: Dietmar Möller/Dobrivoje Popović/Georg Thiele
Modeling, Simulation and Parameter-Estimation of the Human Cardiovascular System

Volume 5: Günter Ludyk
Stability of Time-Variant Discrete-Time Systems

Volume 6: Irmfried Hartmann/Werner Lange/Rainer Poltmann
Robust and Insensitive Design
of Multivariable Feedback Systems
– Multimodel Design –

Irmfried Hartmann
Werner Lange
Rainer Poltmann

Robust and Insensitive Design of Multivariable Feedback Systems – Multimodel Design –

With 53 Figures

Friedr. Vieweg & Sohn Braunschweig / Wiesbaden

CIP-Kurztitelaufnahme der Deutschen Bibliothek

Hartmann, Irmfried:
Robust and insensitive design of multivariable feedback systems – multimodel design / Irmfried Hartmann; Werner Lange; Rainer Poltmann. – Braunschweig; Wiesbaden: Vieweg, 1986.
(Advances in control systems and signal processing; Vol. 6)
ISBN 3-528-08960-1

NE: Lange, Werner:; Poltmann, Rainer:; GT

Editor:

Dr.-Ing. I. Hartmann
Prof. für Regelungstechnik und Systemdynamik
Technische Universität Berlin
Hardenbergstraße 29c
1000 Berlin 12, West Germany

Authors:

Dr.-Ing. I. Hartmann
Prof. für Regelungstechnik und Systemdynamik
Technische Universität Berlin
Hardenbergstraße 29c
1000 Berlin 12, West-Germany

Dipl.-Ing. Werner Lange und
Dipl.-Ing. Rainer Poltmann
Technische Universität
Hardenbergstraße 29c
1000 Berlin 12, West Germany

Druck und buchbinderische Verarbeitung: Lengericher Handelsdruckerei, Lengerich
Printed in Germany

ISSN 0724-9993

ISBN 3-528-08960-1

Preface

The object of this monograph is to present some optimal design methods in a certain sense for obtaining linear multivariable feedback systems such that they become insensitive or robust properties. The treatment is mainly confined to linear time-invariant discrete systems with the exception of the sections 3.1 and 3.3 in which it was used linear time-invariant continuous systems.
Sensitivity and robust methods for multivariable feedback systems have been extensively studied about the last decade and the successful development has led to a much better understanding of the performance of insensitive and robust procedures. The contribution of this monograph is an improved behaviour of the feedback systems under the influence of disturbance, parameter variations and/or nonlinear effects with new design methods. Furthermore, in chapter 3 the reader finds improved necessary and sufficient conditions for robust stability of the closed-loop system.

The authors would like to express their thanks to Mrs. Regina Häde, who typed with great patience our manuscript and Mrs. Monika Thieke who drew the figures and symbols very careful.

Berlin, February 1986

Irmfried Hartmann
Werner Lange
Rainer Poltmann

Contents

Notations

$\mathbb{R}^n$	n-dimensional vector space
$\underline{O}$	zero vector, zero matrix
$\underline{i}_\nu$	ν-th unit vector (chapter 2)
$\underline{I}_n$	n-series unit matrix (chapter 1 and 2)
$\underline{E}_n$	n-series unit matrix (chapter 3)
$adj(\underline{A})$	adjunct matrix of $\underline{A}$
$det(\underline{A})$	determinant of $\underline{A}$

System denotation:

n	system order
r	number of inputs
p	number of outputs
$\underline{G}(s)$	transfer function in the s-domain
$\underline{G}(z)$	discrete transfer function
$\underline{\Delta G}$	difference between two transfer functions
$\Delta(z), N(z)$	characteristic polynomial; denominator polynomial
$Z(z)$	numerator polynomial (in general)
$\underline{K}(s), \underline{F}(s)$	controller transfer function
$\underline{A}$, $\underline{\Phi}$	systemmatrix-continuous,discrete-
$\underline{B}$, $\underline{H}$	input matrix-continuous,discrete-
$\underline{C}$	output matrix
$\underline{D}$	feed through matrix
$\underline{\Delta A}$	difference between two dynamic matrices
$\underline{k}_i$	controller parameter vector
u, $\tilde{u}$	input variable
$\underline{u}$	vector of input-data
$\underline{x}$	state vector
$\underline{y}$	output vector

Nominal command behavior (chapter 2):

$k_i^{(j)}$ elements of a row of the transformed state gain matrix

$\Delta_R(z)$ characteristic polynomial of the nominal command behavior

$\underline{\Gamma}_R(z)$ polynomial matrix designating the nominal command behavior

Controller (chapter 2):

m order of a partial controller

$\Delta_B(z) = z^m + \beta_{m-1} z^{m-1} + \ldots + \beta_1 z + \beta_o$ characteristic polynomial of a partial controller

$k_i^{*(j)}$ free parameters of a partial controller which do not influence the nominal command behavior

Feedback system:

$\underline{S}(z)$ sensitivity function

$\underline{\Gamma}_C(z), \underline{Z}_C(z)$ transfer functions designating the feedback system behavior (chapter 2)

Special notation in chapter 3:

a_i coefficient of denominator-polynomial

$\underline{a}$ vector of denominator polynomial coefficients

$\underline{\tilde{A}}$ sect.3.1.6: dynamic-matrix after an iteration step

b_i coefficient of numerator-polynomial

$\underline{b}$ vector of numerator polynomial coefficients

d sect.3.1: pole-distance
sect.3.3.6: constant in bilinear transformation

dp sect.3.1: distance between poles for permutation

$e(\nu)$ controller input-data

$\underline{H}$ sect.3.1.3: parameter-matrix
sect.3.2.5: input-matrix

m	number of variation-cases
p	complex variable in the p-domain
$s_{i,j}$	controllability-coefficient of pole i and input j
T	sect.3.2: sampling period
$\underline{T}$	sect.3.2: eigenvector-matrix sect.3.3: closed-loop transfer-function
$\underline{U}$	matrix of input-data
$\underline{v}$	vector with poles
$\Delta\underline{V}$	difference between two dynamic-systems
$\underline{W}$	closed-loop perturbation transfer-function
$\underline{w}_p$	vector with permutated poles
$\underline{x}_i$	eigenvector; in sect. 3.2.5 eigenvalues
z	complex variable in the z-domain
$\underline{z}$	vector with powers of z
γ, γ_i	weighting-factor ; in sect.3.2.5 eigenvalues
$\underline{\Gamma}$, $\hat{\underline{\Gamma}}$	weighting-matrices
$\varepsilon(\nu)$	difference-data
ε	sect. 3.3: variation $0 \leq \varepsilon \leq 1$
ϑ	sampling-time
$\underline{\theta}_i$, $\underline{\theta}$	parameter-vector, parameter-matrix
$\underline{\Phi}_c$	dynamic-matrix of controller
$\underline{\Phi}_s$	dynamic-matrix of plant
$\underline{\Phi}_T$	dynamic-matrix, low sampling period
$\underline{\psi}$	data-vector

1 System Theoretical Background

1.1 INTRODUCTION

This book is concerned with the design of insensitive and robust linear time-invariant multivariable feedback systems. Originally, the term robustness of feedback systems was refered to reduce the influence of large plant perturbations where as the term sensitivity concerns the effects of small uncertainties. But in this explanation the sensitivity would be included in the term robustness and really there is hardly a difference.

Our emphasis here, however, is on design techniques that are applicable when either

the plant can be described by a nominal model with known estimation of the parameter variations (model uncertainties)

or

the plant can be represented by several linear time-invariant models with known parameter sets, called multi-model case.

We shall use the term sensitive in the sense of comparison sensitivity. That is the reduction of the sensitivity of the closed loop system to plant perturbations as compared to the sensitivity of the open loop system. The sensitivity function, see eq.(1.32) or eq.(2.57), is given by (time-discret case)

$$\underline{S}(z) = [\underline{I}_p + \underline{L}_p(z)]^{-1} \quad ,$$

also called the inverse return difference matrix. Then the closed loop error $\underline{E}_R(z)$ is related to the open loop error $\underline{E}_s(z)$ by

$$\underline{E}_R(z) = \underline{S}(z)\underline{E}_s(z) \quad .$$

The direct orientation of the design to the comparison sensitivity function makes it possible to minimize the norm of $\underline{S}(z)$ in the lower to middle frequency range with the result of an improved transient response in the present of plant perturbations. In this sense the problem was also dealt with general view by J.B.Cruz et.al [1] and J.C.Doyle et.al.[6]. Robustness will be used here in view of the following definition:

> Given a system property and a discret class of perturbations, then it is to determine a controller such that the system property remains invariant under the influence of perturbations.

That means, we are looking for a controller which is the same for all represented discrete models of the plant and minimizes the difference between the desired and attained closed loop behaviour in the sense of the system property. It is called the multimodel case. Even if there is to regard the sensor-and actuator failure the proposed methods are also applicable on multivariable feedback systems.

In the face of model uncertainties or in the multimodel case it is coming up a stability problem. It is not sufficiently to design a stable feedback system with nominal plant. Either one needs suitable stability bounds including estimated uncertainties or one has to use an appropriate stability criterion. In section 3.3 there are given new necessary and sufficient conditions for robust stability of the closed-loop system. The stability is tested with the help of the principal gains.

In the literature there are also other proposals to find a good robust design but more in the sense of the sensitivity function $\underline{S}(\cdot)$, i.e. A.Dickmann et.al.[5], B.A.Francis [7], A.J.Laub [10] and D.H.Owens et.al.[13].

Now it will be given a short introduction in the background of linear multivariable feedback systems which are needed in this book.

1.2 SOME REPRESENTATION AND PROPERTIES OF LINEAR MULTIVARIABLE SYSTEMS

In this section we give a brief account of various linear models for deterministic dynamical processes. We will begin by describing of time-continuous and time-discrete state-space models with the corresponding transfer functions. Thereafter the different canonical structures of the models with the connected properties are introduced and discussed. An appropriately chosen model structure can often give an insight in the dynamical process and facilitate the design of controller. At the end of the section we will deal with the standard feedback configuration and discuss the meaning of the return difference matrix and its inverse that play an important role in multivariable design.

In this book we consider continuous processes, which can be described (approximately) as finite dimensional linear time-invariant models. In the multivariable case the time-continuous state space models are of the form (without disturbances)

$$\dot{\underline{x}}(t) = \underline{A}\ \underline{x}(t) + \underline{B}\ \underline{u}(t)\ ; \quad \underline{x}(t_o) = \underline{x}_o$$
$$\underline{y}(t) = \underline{C}\ \underline{x}(t) + \underline{D}\ \underline{u}(t)\ , \tag{1.1}$$

wherein the input vector $\underline{u}(t) \in \mathbb{R}^r$, the state vector $\underline{x}(t) \in \mathbb{R}^n$ and the output vector $\underline{y}(t) \in \mathbb{R}^p$. The components $u_i(\cdot)$, $x_i(\cdot)$, $y_i(\cdot) : [0,\infty) \to \mathbb{R}$ are corresponding functions. It is usually to denote

$\underline{A} \in \mathbb{R}^{nxn}$ as system matrix ,

$\underline{B} \in \mathbb{R}^{nxr}$ as input matrix ,

$\underline{C} \in \mathbb{R}^{pxn}$ as output matrix ,

$\underline{D} \in \mathbb{R}^{pxr}$ as distribution matrix .

The dimension of the state vector, n, is called the order of the model.

It will often employ a description as transfer function by design of linear feedback systems. This dynamical input-output relation is derived from eq.(1.1) when the Laplace transformation is used and the initial state $\underline{x}_o$ is setting zero. Under this consideration one finds as transfer function the (pxr) rational matrix

$$\underline{G}(s) = \underline{C}\,[\underline{I}_n s - \underline{A}]^{-1}\,\underline{B} + \underline{D} \qquad (1.2)$$

and the input-output relations holds

$$\underline{Y}(s) = \underline{G}(s)\,\underline{U}(s).$$

We need also the factorization of the rational function $\underline{G}(s)$. There exist two polynomial matrices $\underline{Z}_r(s)$; $\underline{N}_r(s)$ or $\{\underline{Z}_l(s); \underline{N}_l(s)\}$ such that

$$\underline{G}(s) = \underline{Z}_r(s)\underline{N}_r^{-1}(s) = \underline{N}_l^{-1}(s)\underline{Z}_l(s) \qquad (1.3)$$

for all $s \in \mathbb{C}$ and $\{\underline{Z}_r(s); \underline{N}_r(s)\}$ are right coprime ($\{\underline{Z}_l(s); \underline{N}_l(s)\}$ are left coprime), i.e. every greatest common right (left) divisor of $\underline{Z}_r(s)$ ($\underline{Z}_l(s)$) and $\underline{N}_r(s)$ ($\underline{N}_l(s)$) is an unimodular matrix.

The following definition introduces an important property of transfer functions.

Definition 1.1: The rational function $\underline{G}(s)$ is proper (strictly proper) if and only if all elements of $\underline{G}(s)$ are bounded at ∞ (tend to zero at ∞), i.e. $\lim_{s\to\infty} \underline{G}(s) = \underline{G}_\infty$ $(= \underline{O})$.

Remark: Transfer functions derived from state space models (see eq.(1.2)) have the property

$$\lim_{s\to\infty} \underline{G}(s) = \underline{D}\,,$$

i.e. proper or for $\underline{D} = \underline{O}$ strictly proper.

If the state space model $\{\underline{A}, \underline{B}, \underline{C}, \underline{D}\}$ is a minimal realization (def.(1.6) see later), it follows from eqs.(1.2) and (1.3)

$$\det[\underline{I}_n s - A] = g_r \det[\underline{N}_r(s)] = g_l \det[\underline{N}_l(s)] \qquad (1.4)$$

with g_r, $g_l \in \mathbb{R}$ is constant and all zeros of eq.(1.4) are poles of the transfer function $\underline{G}(s)$. The zeros of $\underline{G}(s)$ are determined from the Rosenbrock matrix

$$\underline{P}(s) := \begin{bmatrix} \underline{I}_n s - \underline{A} & \underline{B} \\ \underline{C} & \underline{D} \end{bmatrix} \begin{matrix} n \\ p \end{matrix} \quad \text{(columns: } n, r\text{)} \quad . \qquad (1.5)$$

β_ν is a zero of $\underline{G}(s)$ if and only if the condition

$$\mathrm{rk}[\underline{P}(\beta_\nu)] < n + \min(p,r) \qquad (1.6)$$

is satisfied. More details about rational matrices and the meaning of zeros can be found in C.A.Desoer et.al.[3], in T.Kailath [8] and in H.W.Knobloch et.al [9].

A time-discrete model can derive from a linear time-continuous system under discrete control. Assume that the input is generated by a zero-order hold then it is with T as sampling interval

$$\underline{u}(\nu) := \underline{u}(\nu T) \qquad \text{for} \quad \nu T \leq t < (\nu+1)T \ .$$

The output and state-space vector samples of the time-continuous system (1.1) then satisfy a linear time-discret state space model of the form

$$\begin{aligned} \underline{x}(\nu+1) &= \underline{\Phi}\ \underline{x}(\nu) + \underline{H}\ \underline{u}(\nu) \\ \underline{y}(\nu) &= \underline{C}\ \underline{x}(\nu) + \underline{D}\ \underline{u}(\nu) \end{aligned} \qquad (1.7)$$

with $\underline{x}(\nu) := \underline{x}(\nu T)$ and $\underline{y}(\nu) := \underline{y}(\nu T)$. The relationship between $\underline{A}$, $\underline{B}$ and $\underline{\Phi}$, $\underline{H}$ may be expressed as

$$\underline{\Phi} := \underline{\Phi}(T) = e^{\underline{A}T} = \sum_{i=o}^{\infty} \frac{\underline{A}^i\ T^i}{i!}$$

$$\underline{H} = \int_o^T \underline{\Phi}(\tau)\underline{B}\ d\tau = \sum_{i=o}^{\infty} \frac{\underline{A}^i\ T^{i+1}}{(i+1)!}\ \underline{B} \ .$$

There are four significant properties of the state space model (1.7) and in the time-continuous case (eq.1.1) one defines the properties in a simalar way.
The properties are called controllable, reachable, observable and reconstructible.

Definition 1.2: A state vector $\underline{x}^*$ of the system (1.7) is said to be

controllable if there exists a finite N and a control sequence $\{\underline{u}(0),\ldots,\underline{u}(N-1)\}$ which drives the initial state $\underline{x}(0) = \underline{x}^*$ to $\underline{x}(N) = \underline{0}$. The system (1.7) is completely controllable if every state is controllable.

reachable if there exists a finite N and a control sequence $\{\underline{u}(0),\ldots,\underline{u}(N-1)\}$ which drives the initial state $\underline{x}(0) = \underline{0}$ to $\underline{x}(N) = \underline{x}^*$.
The system (1.7) is completely reachable if every state is reachable.

observable if there exists a finite N such that the initial state $\underline{x}(0) = \underline{x}^*$ can be uniquely determined from the output sequence $\{\underline{y}(0),\ldots,\underline{y}(N-1)\}$ for each allowable input sequence. The system (1.7) is completely observable if every state $\underline{x}^*$ is observable.

reconstructible if there exists a finite N such that the state vector $\underline{x}(0) = \underline{x}^*$ can be uniquely determined from the output sequence $\{\underline{y}(-N+1),\ldots,y(0)\}$ for each allowable input sequence.
The system (1.7) is completely reconstructible if every state $\underline{x}^*$ is reconstructible.

In order to examine these properties we introduce the so called controllability matrix

$$\underline{W} = [\underline{H},\ \underline{\Phi}\ \underline{H},\ \ldots,\ \underline{\Phi}^{n-1}\underline{H}] \tag{1.8}$$

and the observability matrix

$$\underline{M} = [\underline{C}',\ \underline{\Phi}'\underline{C}',\ldots,\underline{\Phi}'^{n-1}\underline{C}']\ . \tag{1.9}$$

Then it follows the known result.

Lemma 1.3: The state space model (1.7) of order n is

completely controllable if $rk[\underline{W}] = n$ and the condition is necessary if $\underline{\Phi}$ is nonsingular,
completely reachable if and only if $rk[\underline{W}] = n$,

completely observable if and only if $rk[\underline{M}] = n$,

completely reconstructible if $rk[\underline{M}] = n$ and the condition is necessary if $\underline{\Phi}$ is nonsingular.

Proof: see T.Kailath [8].

Canonical forms of the state space model

Equivalent representations to eq.(1.7) can be obtained by choosing a new basis for the state space. That can be done with a nonsingular transformation of the state vector $\underline{x}(\nu)$ such that

$$\tilde{\underline{x}}(\nu) = \underline{T}\ \underline{x}(\nu) \qquad (\underline{T} \text{ nonsingular})$$

is the new state vector in the transformed state space model

$$\begin{aligned} \tilde{\underline{x}}(\nu+1) &= \tilde{\underline{\Phi}}\ \tilde{\underline{x}}(\nu) + \tilde{\underline{H}}\ \underline{u}(\nu) \qquad \tilde{\underline{x}}(0) = \tilde{\underline{x}}_o \\ \underline{y}(\nu) &= \tilde{\underline{C}}\ \tilde{\underline{x}}(\nu) + \underline{D}\ \underline{u}(\nu) \end{aligned} \qquad (1.10)$$

where $\tilde{\underline{\Phi}} = \underline{T}\ \underline{\Phi}\ \underline{T}^{-1}$, $\tilde{\underline{H}} = \underline{T}\ \underline{H}$ and $\tilde{\underline{C}} = \underline{C}\ \underline{T}^{-1}$. Each nonsingular transformation leads us to a new equivalent representation (1.10). Under this consideration we can choose special forms for the transformed state space models which are convenient to use for particular applications. On the other hand it is important to know which properties of the state space model are unchangeable under nonsingular transformations.
In this sense it is useable for the multivariable case to define a new property that is connected with the controllability.

Definition 1.4: There are given the matrices $(\underline{\Phi}, \underline{H})$ with $\underline{H} = [\underline{h}_1,\dots,\underline{h}_r]$ and $rk[\underline{H}] = r$.
Now it may be constructed a new matrix $\underline{W}_o$ by simply searching the smallest integers μ_i in the controllability matrix $\underline{W}$, eq.(1.8), from left to right such that the column vectors $\underline{\Phi}^{\mu_i}\ \underline{h}_i$ $(i = 1,\dots,r)$ are linear dependent of the previous column vectors of $\underline{W}$. This procedure gives us a set of linear independent columns and generates the matrix $\underline{W}_o$.

The integers μ_i (i = 1,...,r) of the matrices ($\underline{\Phi}$, $\underline{H}$) are said to be controllability indices and the largest value

$$\hat{\mu} = \max_i (\mu_i) \tag{1.11}$$

is called the maximum controllability index. The resulting columns of $\underline{W}_o$ are then rearrenged into the following form

$$\underline{W}_o = [\underline{h}_1,\ldots,\underline{\Phi}^{\mu_1 - 1}\underline{h}_1, \underline{h}_2,\ldots,\underline{h}_r,\ldots,\underline{\Phi}^{\mu_r - 1}\underline{h}_r]. \tag{1.12}$$

If rk[$\underline{W}$] = n, then it follows

$$\sum_{i=1}^{r} \mu_i = n \quad .$$

Furthermore one gets for the $\hat{\mu}$ the bounds

$$\frac{n}{r} \leq \hat{\mu} \leq (n-r+1) \quad .$$

The proof is easy to see and the reader should verify the statement.
In the same manner one defines the observability indices for the matrices ($\underline{\Phi}$, $\underline{C}$) with regard to eq.(1.9).

Theorem 1.5: It is given a completely controllable state space model with matrices ($\underline{\Phi}$, $\underline{H}$) and rk[$\underline{H}$] = r. Then it will be considered a feedback with arbitrary state space controller $\underline{K}$, the arbitrary nonsingular transformations in the state space and input space, i.e.

$$(\underline{\Phi}, \underline{H}) \rightarrow ([\underline{\Phi} - \underline{H}\,\underline{K}], \underline{H}) \tag{1.13}$$

$$(\underline{\Phi}, \underline{H}) \rightarrow (\underline{T}\,\underline{\Phi}\,\underline{T}^{-1}, \underline{T}\,\underline{H}) \tag{1.14}$$

$$(\underline{\Phi}, \underline{H}) \rightarrow (\underline{\Phi}, \underline{H}\,\underline{Q}) \quad . \tag{1.15}$$

The unarranged set of controllability indices, given of ($\underline{\Phi}$, $\underline{H}$), is invariant under the transformations $\underline{K}$, $\underline{T}$ and $\underline{Q}$.

Furthermore the controllability indices imply an unique canonical form of the state space model

$$\underline{\tilde{x}}(\nu+1) = \begin{bmatrix} \underline{J}_1 & & \underline{0} \\ \underline{0} & \ddots & \underline{0} \\ & \underline{0} & \underline{J}_r \end{bmatrix} \underline{\tilde{x}}(\nu) + [\underline{e}_1,\ldots,\underline{e}_r]\ \underline{\tilde{u}}(\nu) \quad (1.16)$$

with

$$\underline{J}_i = \begin{bmatrix} 0 & 1 & & \underline{0} \\ \vdots & & \ddots & 1 \\ 0 & \cdots & \cdots & 0 \end{bmatrix} \quad \text{as } (\mu_i \times \mu_i) \text{ Jordanmatrix (eigenvalue 0)}$$

and

$$\underline{e}_i' = [0 \ \ldots \ 1 \ \ldots \ 0] \quad (1.17)$$

$$\uparrow \text{ at position } \sum_{j=1}^{i} \mu_j \ .$$

At first P.Brunowsky [2] and V.M.Popov [14] derived this theorem. Here we give only the essential points of this proof. Under the assumption rk[$\underline{T}$] = n and rk[$\underline{Q}$] = r the assertions will be fulfilled to relations (1.13) til (1.15). It follows from rk[$\underline{Q}$] = r that

$$\text{rk}[\underline{H}] = \text{rk}[\underline{H}\ \underline{Q}] \ .$$

One gets the controllability matrix of the transformed state space model

$$[\underline{H}\ \underline{Q},\ \underline{\Phi}\ \underline{H}\ \underline{Q},\ldots]$$

and the matrix $\underline{Q}$ generates r linear combinations of the r column vectors of $\underline{H}$ such that the values of the controllability indices do not change.
Under the nonsingular transformation $\underline{T}$, eq.(1.14), the controllability matrix $\underline{W}_o$, eq.(1.12), is changing in $\underline{T}\ \underline{W}_o$ but the controllability indices remain invariant. It follows from the fact, that each basis $\{\underline{h}_i,\ldots,\underline{\Phi}^{\mu_i-1}\underline{h}_i\}$ generates a subspace of dimension μ_i and the whole space (range of $\underline{W}_o$) has the dimension $n = \sum_{i=1}^{r} \mu_i$. But a nonsingular map does not change the dimensions μ_i (i = 1,...r) and n.

The controllability matrix, that belongs to eq.(1.13), can be expressed in the form ($\underline{\Phi}_o := [\underline{\Phi} - \underline{H}\ \underline{K}]$)

$$[\underline{H}, \underline{\Phi}\ \underline{H}, \ldots] = [\underline{H}, \underline{\Phi}_o \underline{H}, \ldots] \begin{bmatrix} \underline{I}_r & \underline{K}\ \underline{H} & \cdots & \underline{K}\ \underline{\Phi}^{n-2}\ \underline{H} \\ \underline{O} & \underline{I}_r & \ddots & \vdots \\ \vdots & & \underline{I}_r & \underline{K}\ \underline{H} \\ \underline{O} & \cdots & \underline{O} & \underline{I}_r \end{bmatrix} .$$

The reader should verify this equation for n = 3. On the right side the triangle matrix has always the full rank. That means the structures of both controllability matrices are the same. The figure 1.1 shows us one aspect of the invariant controllability indices.

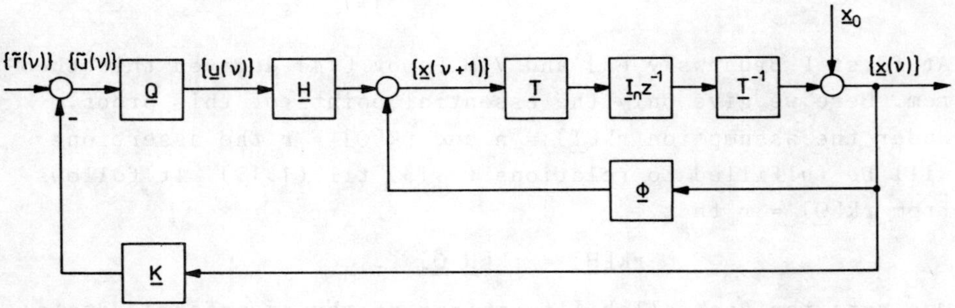

Fig. 1.1: Transformed feedback system with the same controllability indices as $(\underline{\Phi}, \underline{H})$.

In the proof the next step is to verify the eq.(1.16). In this part we derive two canonical forms of the state space model $(\underline{\Phi}, \underline{H})$.

The following (nxn)-matrix

$$\underline{T}_c := \begin{bmatrix} \underline{t}'_1 \\ \vdots \\ \underline{t}'_1 \underline{\Phi}^{\mu_1 - 1} \\ \hline \underline{t}'_2 \\ \vdots \\ \underline{t}'_r \underline{\Phi}^{\mu_r - 1} \end{bmatrix} \qquad (1.18)$$

can be used with the condition

$$\underline{t}_i^{'} \, \underline{W}_o = \underline{e}_i^{'} \qquad i = 1,\dots,r \qquad (1.19)$$

($\underline{W}_o$, $\underline{e}_i$ see eq.'s (1.12) and (1.17)) as transformation. But at first we recognize the following relation:

- In regard to eq.(1.19) one gets the statement

$$\underline{t}_i^{'} \, \underline{\Phi}^m \underline{h}_\nu = \begin{cases} 0 & \text{otherwise} \\ 1 & \nu = i \text{ and } m = \mu_i - 1 \end{cases}$$

  $\nu, i = 1 \dots r;\; m < \mu_\nu$.

- Each vector $\underline{t}_i$ has n components and under the conditions (1.19) and $\text{rk}[\underline{W}_o] = n$ one gets n equations for each $\underline{t}_i$ such that

$$\underline{t}_i^{'} = \underline{e}_i^{'} \underline{W}_o^{-1} \qquad i = 1,\dots,r$$

  is uniquely determined.

Now the transformation can be accomplished. It holds for the transforming state space model

$$\tilde{\underline{x}}(\nu+1) = \underline{T}_c \underline{\Phi} \, \underline{x}(\nu) + \underline{T}_c \underline{H} \, \underline{u}(\nu) \; .$$

In detail one obtains for the transformed system matrix

$$\underline{T}_c \underline{\Phi} \, \underline{x}(\nu) = \begin{bmatrix} \underline{t}_1^{'} \underline{\Phi} \, \underline{x}(\nu) \\ \vdots \\ \underline{t}_1^{'} \underline{\Phi}^{\mu_1} \underline{x}(\nu) \\ \hline \underline{t}_2^{'} \underline{\Phi} \, \underline{x}(\nu) \\ \vdots \\ \hline \underline{t}_r^{'} \underline{\Phi}^{\mu_r} \underline{x}(\nu) \end{bmatrix} = \begin{bmatrix} \tilde{x}_2(\nu) \\ \vdots \\ \tilde{x}_{\mu_1}(\nu) \\ \underline{\alpha}_1^{'} \tilde{\underline{x}}(\nu) \\ \hline \tilde{x}_{\mu_1+2}(\nu) \\ \vdots \\ \hline \underline{\alpha}_r^{'} \, \tilde{\underline{x}}(\nu) \end{bmatrix} = \underline{T}_c \underline{\Phi} \, \underline{T}_c^{-1} \tilde{\underline{x}}(\nu) = \tilde{\underline{\Phi}} \, \tilde{\underline{x}}(\nu)$$

with $\underline{\alpha}_i^{'} = [\underline{\alpha}_{1i}^{'}, \dots, \underline{\alpha}_{ri}^{'}]$ and $\underline{\alpha}_{ji} \in \mathbb{R}^{\mu_j}$ $i = 1,\dots,r$. In regard to $\{u(\nu)\}$ the input can be expressed by

$$\underline{T}_c\underline{H} = \left[\begin{array}{c|c|c} \underline{t}_1'\underline{h}_1 & & \underline{t}_1'\underline{h}_r \\ \vdots & & \vdots \\ \underline{t}_1'\underline{\Phi}^{\mu_1-1}\underline{h}_1 & \cdots & \underline{t}_1'\underline{\Phi}^{\mu_1-1}\underline{h}_r \\ \hline \vdots & & \vdots \\ \underline{t}_r'\underline{\Phi}^{\mu_r-1}\underline{h}_1 & \cdots & \underline{t}_r'\underline{\Phi}^{\mu_r-1}\underline{h}_r \end{array}\right] = [\underline{h}_{c1},\ldots,\underline{h}_{cr}]$$

and under the conditions $\mu_1 \geq \mu_2 \geq \ldots \geq \mu_r$ and eq.(1.19) it can be verified for the elements of $\underline{T}_c\underline{H}$

$$\underline{t}_j\underline{\Phi}^{\nu}\underline{h}_l \begin{cases} = 0 & j \neq l \quad \nu \leq (\mu_l-1) \\ = 1 & j = l \quad \nu = (\mu_l-1) \qquad l = 1,\ldots,r \\ = 0 & j = l \quad \nu < (\mu_l-1) \end{cases}$$

and

$$\underline{t}_j\underline{\Phi}^{\varkappa}\underline{h}_l \begin{cases} \neq 0 & \text{(in general)} \ \varkappa = (\mu_j-1) \\ & \qquad j = 1,\ldots,(l-1) \\ = 0 & \text{elsewhere for } \varkappa < (\mu_j-1) \end{cases} .$$

Now the column vectors $\underline{h}_{ci}$ have only i nonzero components in the positions

$$\mu_1,\ (\mu_1+\mu_2),\ldots,\ \sum_{j=1}^{i}\mu_j \ ,$$

e.g. $\underline{h}_{c1} = \underline{e}_1$. In order to get the input matrix in eq.(1.16) we transform with the triangular matrix $\underline{Q}$, see fig.1.1.

$$\tilde{\underline{H}} = \underline{T}_c\underline{H}\ \underline{Q} = \underline{T}_c\underline{H} \begin{bmatrix} 1 & q_{12} & \cdots & & q_{1r} \\ 0 & 1 & & & \vdots \\ \vdots & 0 & \cdot & & \vdots \\ \vdots & \vdots & & \cdot & q_{r-1,r} \\ 0 & 0 & \cdots & 0 & 1 \end{bmatrix}$$

$$= \left[\underline{e}_1, [q_{12}\underline{e}_1 + \underline{h}_{c2}], \ldots, \sum_{i=1}^{r} q_{ir}\,\underline{h}_{ci}\right]$$

with $q_{rr} = 1$ and $\underline{h}_{c1} = \underline{e}_1$. The coefficients q_{ij} can be computed from the equations

$$q_{12}\underline{e}_1 + \underline{h}_{c2} = \underline{e}_2, \ldots, \sum_{i=1}^{r} q_{ir}\underline{h}_{ci} = \underline{e}_r$$

because the $q_{\nu j}\ \underline{h}_{c\nu}$ compensates the $\sum_{i=1}^{\nu}\mu_i$ component of

$\sum_{i=\nu}^{j} q_{ij}\underline{h}_{ci}$ with $\nu = (r-1),\ldots,1$ and $j = r,\ldots,2$.

Now the complete state space model takes the following canonical matrix form:

$$\tilde{\underline{x}}(\nu+1) = \underbrace{\left[\begin{array}{c|c|c}
\begin{matrix}0&1&0&\cdots&0\\ \vdots&&\ddots&&\vdots\\ \vdots&&&\ddots&0\\ 0&&&&1\\ &&\underline{\alpha}'_{11}&&\end{matrix} & \begin{matrix}\underline{0}\\ \\ \cdots\end{matrix} & \begin{matrix}\underline{0}\\ \\ \underline{\alpha}'_{r1}\end{matrix} \\ \hline
\vdots & & \vdots \\ \hline
\begin{matrix}\underline{0}\\ \\ \underline{\alpha}'_{1r}\end{matrix} & \begin{matrix}\underline{0}\\ \\ \cdots\end{matrix} & \begin{matrix}0&1&0&\cdots&0\\ \vdots&&\ddots&&\vdots\\ \vdots&&&\ddots&0\\ 0&&&&1\\ &&\underline{\alpha}'_{rr}&&\end{matrix}
\end{array}\right]}_{\mu_1 \;\cdots\; \mu_r\text{ columns}}
\begin{matrix}\left.\right\}\mu_1\text{rows}\\ \\ \\ \left.\right\}\mu_r\text{rows}\end{matrix}
\tilde{\underline{x}}(\nu) + \left[\begin{array}{c}
\begin{matrix}0&0&\cdots&0\\ \vdots&\vdots&&\vdots\\ 0&\cdot&&\cdot\\ 1&0&\cdots&0\end{matrix}\\ \hline
\vdots\\ \hline
\begin{matrix}0&\cdot&\cdot&0\\ \vdots&&&\vdots\\ \cdot&&&0\\ 0&\cdots&0&1\end{matrix}
\end{array}\right]\tilde{\underline{u}}(\nu)$$

$$= \tilde{\underline{\Phi}}\,\tilde{\underline{x}}(\nu) + \tilde{\underline{H}}\,\tilde{\underline{u}}(\nu) \qquad \textbf{(1.20)}$$

This is called the controllability form. With a suitable transformation one can find other canonical forms. In chapter 2, eq.'s (2.20) til (2.23), there will be applied the observable canonical form that can be derived in the similar way as above.

Now it is easy to show with the feedback

$$\tilde{\underline{u}}(\nu) = -\tilde{\underline{K}}\,\tilde{\underline{x}}(\nu) + \tilde{\underline{r}}(\nu)$$

and the steady state controller

$$\tilde{\underline{K}} = \begin{bmatrix}\tilde{\underline{k}}'_1\\ \vdots\\ \tilde{\underline{k}}'_r\end{bmatrix} = \begin{bmatrix}\underline{\alpha}'_{11} & \cdots & \underline{\alpha}'_{r1}\\ \\ \underline{\alpha}'_{r1} & \cdots & \underline{\alpha}'_{rr}\end{bmatrix} \qquad \textbf{(1.21)}$$

(with μ_1- … μ_r columns)

that the canonical form (1.20) is changing into eq.(1.16).
It holds

$$\tilde{\underline{x}}(\nu+1) = [\tilde{\underline{\Phi}} - \tilde{\underline{H}}\,\tilde{\underline{K}}]\,\tilde{\underline{x}}(\nu) + \tilde{\underline{H}}\,\tilde{\underline{r}}(\nu)$$

and

$$\underline{\tilde{H}}\;\underline{\tilde{K}} = \begin{bmatrix} \underline{0}' \\ \vdots \\ \underline{\tilde{k}}_1' \\ \hline \vdots \\ \hline \underline{0}' \\ \vdots \\ \underline{\tilde{k}}_r' \end{bmatrix} \begin{matrix} \left.\vphantom{\begin{matrix}a\\b\\c\end{matrix}}\right\} \mu_1 \text{ rows} \\ \\ \left.\vphantom{\begin{matrix}a\\b\\c\end{matrix}}\right\} \mu_r \text{ rows} \end{matrix} \; .$$

The matrix $\underline{\Phi}_o = [\underline{\tilde{\Phi}} - \underline{\tilde{H}}\;\underline{\tilde{K}}]$ takes now the form of the system matrix in eq.(1.16).

The reader should verify the following properties of the matrix $\underline{\Phi}_o$. It is valid, see eq.(1.17),

$$\underline{\Phi}_o^{\mu_i}\,\underline{e}_i = \underline{0} \qquad \text{and} \qquad \underline{\Phi}_o^{\hat{\mu}}\,\underline{x}_o = \underline{0} \quad \text{for each } \underline{x}_o \in \mathbb{R}^n .$$

Input-output-relation

In the preceding explanations we discussed canonical models. Under the transformation (1.14) the models were shown to be input-output equivalent to the original system for each initial state. Now we restrict our attention to the completely controllable and completely observable part of the system.

Definition 1.6: A completely controllable and observable state space model (1.7) is said to be a minimal realization.

In the minimal model there is no possibility to reduce the state dimension. Note that the minimal model does not describe the uncontrollable and unobservable part of the system.

Now it can be derived a time-discrete input-output relation from eq.(1.7). When the Z-transformation is applied and the initial state $\underline{x}_o$ is setting zero one finds as Z-transfer function the (pxr) rational matrix

$$\underline{G}(z) = \underline{C}[\underline{I}_n z - \underline{\Phi}]^{-1}\,\underline{H} + \underline{D} \qquad (1.22)$$

and the input-output relation holds

$$\underline{Y}(z) = \underline{G}(z)\underline{U}(z) \; .$$

The factorization of the rational function $\underline{G}(z)$ and the determination of poles and zeros are the same as in the continuous case, see eq.'s (1.3) til (1.6). Also the property (strictly) proper is given by

$$\lim_{z\to\infty} \underline{G}(z) = \underline{D} \quad (\underline{0}) .$$

Singulare values of a matrix

The eigenvalues of a square matrix $\underline{A}$ characterize together with the eigenvectors the properties of the matrix. The gain of the matrix $\underline{A}$ is given by the eigenvalue in the direction of the appropiate eigenvector. One finds the eigenvalues λ_ν and eigenvectors $\underline{\varphi}_\nu$ by solving the equations

$$\det[\underline{I}_n \lambda - \underline{A}] = 0$$

and

$$\underline{A}\,\underline{\varphi}_\nu = \lambda_\nu\,\underline{\varphi}_\nu \quad .$$

When we have to judge a non-square matrix, e.g. a (pxr) rational matrix $\underline{G}(z)$ with fixed z, then one needs the concept of singular values. The singular values allows one to determine the smallest and greatest gain of any matrix considering all directions.
If $\underline{A}$ is any (pxr) matrix (with real or complex elements) then the smallest and greatest singular value are defined as follows:

$$\bar{\sigma}[\underline{A}] := \max_{||\underline{x}||=1} ||\underline{A}\,\underline{x}|| = \sqrt{\lambda_{max}[\underline{A}^*\underline{A}]} \qquad (1.23)$$

$$\underline{\sigma}[\underline{A}] := \min_{||\underline{x}||=1} ||\underline{A}\,\underline{x}|| = \sqrt{\lambda_{min}[\underline{A}^*\underline{A}]} \qquad (1.24)$$

where $||\cdot||$ is the usual Euclidean norm and $[\cdot]^*$ denotes conjugate transpose. Now some elementary properties of singular values are stated.

- If $\underline{A}$ is square and $\underline{\sigma}[\underline{A}] > 0$ then $\underline{A}^{-1}$ exists and

$$\underline{\sigma}[\underline{A}] = \frac{1}{\bar{\sigma}[\underline{A}^{-1}]} \quad . \qquad (1.25)$$

- If $\lambda_\nu[\underline{A}]$ any eigenvalue of a square matrix $\underline{A}$ then it is valid the bounding relations

$$\underline{\sigma}[\underline{A}] \leq \lambda_\nu[\underline{A}] \leq \bar{\sigma}[\underline{A}] \quad . \tag{1.26}$$

- Let $\underline{A}$ a (pxr)-matrix and $\underline{B}$ a (rxq)-matrix then it holds

$$\bar{\sigma}[\underline{A}\ \underline{B}] \leq \bar{\sigma}[\underline{A}]\bar{\sigma}[\underline{B}]$$

and (1.27)

$$\underline{\sigma}[\underline{A}]\underline{\sigma}[\underline{B}] \leq \underline{\sigma}[\underline{A}\ \underline{B}] \ .$$

- Let $\underline{A}$ and $\underline{\Delta A}$ (pxr)-matrices then it can be shown under the assumption $\underline{\sigma}[\underline{A}] > \bar{\sigma}[\underline{\Delta A}]$

$$\underline{\sigma}[\underline{A} + \underline{\Delta A}] \geq \underline{\sigma}[\underline{A}] - \bar{\sigma}[\underline{\Delta A}] \quad ,$$

(1.28)

$$|\underline{\sigma}[\underline{A} + \underline{\Delta A}] - \underline{\sigma}(\underline{A})| \leq \bar{\sigma}[\underline{\Delta A}]$$

and for p = r

$$\underline{\sigma}[\underline{A}] - 1 \leq \underline{\sigma}[\underline{I} + \underline{A}] \leq \underline{\sigma}[\underline{A}] + 1 \ . \tag{1.29}$$

In order to determine alle positive singular values σ_ν of a (pxr)-matrix $\underline{A}$ we have to solve the equation

$$\det \underset{\quad p \qquad r}{\begin{bmatrix} \sigma\underline{I}_p & \underline{A} \\ \underline{A}^* & \sigma\underline{I}_r \end{bmatrix}} \begin{matrix} p \\ \\ r \end{matrix} = \sigma^{(p-r)}\det[\sigma^2\underline{I}_r - \underline{A}^*\underline{A}] = \sigma^{(r-p)}\det[\sigma^2\underline{I}_p - \underline{A}\ \underline{A}^*] \overset{!}{=} 0 \tag{1.30}$$

$\underline{A}^*\underline{A}$ and $\underline{A}\ \underline{A}^*$ have the same number of nonzero eigenvalues $\sigma_\nu^2>0$ and these σ_ν are called the (positive) singular values.

feedback systems

Now it will deal with the standard feedback configuration shown in fig.1.2. It consists of the plant, the forward and backward controller ($\underline{K}$, $\underline{F}$) forced by commands $\tilde{R}$, disturbance Ξ_1 and measurement noise $\underline{\Xi}_2$. The transfer matrices $\underline{G}$, $\underline{K}$ and $\underline{F}$

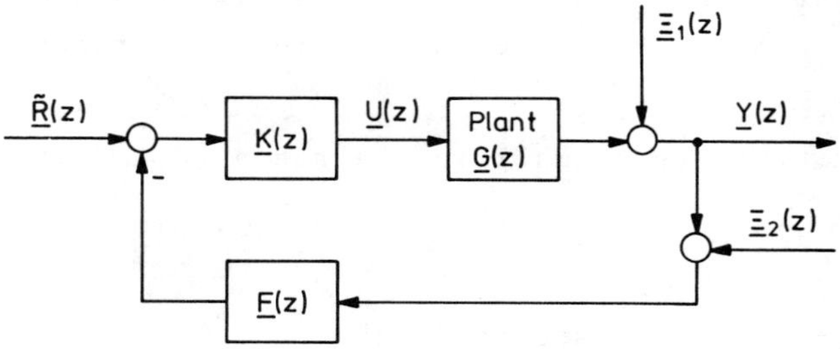

Fig.1.2: Multivariable feedback system with $\underline{Y}, \tilde{\underline{R}} \in \mathbb{R}^p$ and $\underline{u} \in \mathbb{R}^r$.

have appropriate rows and columns such that the open loop transfer matrix

$$\underline{L}_p(z) = \underline{G}(z)\underline{K}(z)\underline{F}(z) \tag{1.31}$$

exists and is (pxp) square.
The feedback system is described by the following input-output-relations (see fig.1.2):

$$\begin{aligned}\underline{Y}(z) &= [\underline{I}_p + \underline{L}_p(z)]^{-1}\underline{G}(z)\underline{K}(z)\tilde{\underline{R}}(z) \\ &\quad + [\underline{I}_p + \underline{L}_p(z)]^{-1}[\underline{\Xi}_1(z) - \underline{L}_p(z)\underline{\Xi}_2(z)] \\ &= \underline{T}_R(z)\tilde{\underline{R}}(z) + \underline{S}(z)[\underline{\Xi}_1(z) - \underline{L}_p(z)\underline{\Xi}_2(z)]\end{aligned} \tag{1.32}$$

with $\tilde{\underline{L}}_r(z) = \underline{K}(z)\underline{F}(z)\underline{G}(z)$ one gets for the input of the plant

$$\begin{aligned}\underline{U}(z) &= [\underline{I}_r + \tilde{\underline{L}}_r(z)]^{-1}\underline{K}(z)\tilde{\underline{R}}(z) \\ &\quad - [\underline{I}_r + \tilde{\underline{L}}_r(z)]^{-1}\underline{K}(z)\underline{F}(z)[\underline{\Xi}_2(z) + \underline{\Xi}_1(z)].\end{aligned} \tag{1.33}$$

Under the assumption that $\underline{F}(z)$ is invertible it holds

$$[\underline{T}_R(z) - \underline{F}^{-1}(z)] = -[\underline{I}_p + \underline{L}_p(z)]^{-1}\underline{F}^{-1}(z). \tag{1.34}$$

The expressions $[\underline{I}_p + \underline{L}_p]$ and $\underline{S} = [\underline{I}_p + \underline{L}_p]^{-1}$ are called the <u>return difference matrix</u> and <u>sensitivity function</u>.

Now we consider the smallest and greatest singular values of the frequency matrices $\underline{T}_R$ and $\underline{S}$ ($z \rightarrow e^{j\omega T}$). It is assumed that the feedback system is effective in the frequency range $[0,\alpha_B]$ with $\alpha_B := \omega_B T$.

If the sensitivity is very small in the range $[0,\alpha_B]$ it follows:

$$||\underline{S}(e^{j\alpha})|| = \bar{\sigma}[\underline{S}] << 1$$

$$\leftrightarrow \quad \underline{\sigma}[\underline{I}_p + \underline{L}_p] >> 1 \text{ , see eq.(1.25)} \tag{1.35}$$

$$\leftrightarrow \quad \underline{\sigma}[\underline{G}\ \underline{K}\ \underline{F}] >> 1 \text{ , see eq.(1.29)} \tag{1.36}$$

for all $\alpha \in [0,\alpha_B]$. In this case equation (1.32) can be approximately described by

$$\underline{Y} \approx \underline{F}^{-1}\ \tilde{\underline{R}} - \underline{\Xi}_2 \quad .$$

That means large loop gains, eq.(1.36), can make the control system with measurement noise $\underline{\Xi}_2$ uneffective. From the eq. (1.34) it follows

$$||\underline{T}_R - \underline{F}^{-1}|| << ||\underline{F}^{-1}|| \quad . \tag{1.37}$$

Assume, that the true plant can be described by $\underline{G}(z)$ and the nominal model $\underline{G}_o(z)$ was used in the design. For instance, the uncertainty can be determined by

$$\Delta\underline{G}(z) = \underline{G}(z) - \underline{G}_o(z) \ .$$

$\Delta\underline{G}(z)$ may be caused by parameter changes, by neglected dynamics or by other unspecified effects.

After it will replace the r = p input-output plant $\underline{G}(z)$ in the eq.'s (1.31) and (1.32) one gets the following expressions

$$\underline{L}_p(z) = \underline{G}_o(z)\underline{K}(z)\underline{F}(z) + \underline{\Delta G}(z)\underline{K}(z)\underline{F}(z)$$

$$= \underline{L}_o(z) + \underline{\Delta L}(z)$$

and with $\underline{\Xi}_1(z) = \underline{\Xi}_2(z) \equiv 0$

$$\underline{Y}(z) = [\underline{I}_p+\underline{L}_o(z) + \underline{\Delta L}(z)]^{-1}[\underline{G}_o(z) + \Delta\underline{G}(z)]\underline{K}\ \tilde{\underline{R}}(z)$$

$$= [\underline{I}_p+[\underline{I}_p+\underline{L}_o]^{-1}\underline{\Delta L}]^{-1}[\underline{I}_p+\underline{L}_o]^{-1}[\underline{G}_o+\Delta\underline{G}]\underline{K}\ \tilde{\underline{R}}(z) \ .$$

Assume, that the (pxp) matrix $\underline{G}_o(z)$ is invertible. The reference behaviour is approximately the same under the influence of uncertainties $\Delta\underline{G}(z)$, if the greatest singular value of $\underline{S}_o = [\underline{I}_p+\underline{L}_o]^{-1}$, $\underline{\Delta L}$ and $[\underline{G}_o^{-1}\underline{\Delta G}]$ are satisfied the conditions

$$\bar{\sigma}[\underline{S}_o] << \frac{1}{\bar{\sigma}[\underline{\Delta L}]} \qquad \textbf{(1.38)}$$

and

$$\bar{\sigma}[\underline{G}_o^{-1}\underline{\Delta G}] \ << 1 \ . \qquad \textbf{(1.39)}$$

In the second chapter the conditions (1.35) and (1.38) will be minimized in the sense of the Euclidean norm.
The following identities are usefully to derive some relations. It is easy to show that the expression

$$\underline{T}_R(z) = [\underline{I}_p+\underline{G}\ \underline{K}\ \underline{F}]^{-1}\underline{G}\ \underline{K} = \underline{G}\ \underline{K}[\underline{I}_p+\underline{F}\ \underline{G}\ \underline{K}]^{-1} \qquad \textbf{(1.40)}$$

is valid. To verify this equation we apply the identities

$$\underline{G}\ \underline{K} + \underline{G}\ \underline{K}\ \underline{F}\ \underline{G}\ \underline{K} = [\underline{I}_p+ \underline{G}\ \underline{K}\ \underline{F}]\ \underline{G}\ \underline{K}$$

$$= \underline{G}\ \underline{K}[\underline{I}_p+\underline{F}\ \underline{G}\ \underline{K}] \ .$$

If we set equal the right side of these equations it follows eq.(1.40).
Assume that the forward transfer function $\underline{G}_k(z) := \underline{G}(z)\underline{K}(z)$ has the minimal realization $\{\underline{\Phi}, \underline{H}, \underline{C}, \underline{D}\}$ and in fig.1.2 $\underline{F}(z)$ is a constant matrix then it holds

$$\underline{G}_k(z) = \underline{C}[\underline{I}_n z - \underline{\Phi}]^{-1}\ \underline{H} + \underline{D}$$

and the state space model of the feedback system takes the form with $\tilde{\underline{u}}(\nu) = \tilde{\underline{r}}(\nu) - \underline{F}\ \underline{y}(\nu)$

$$\underline{x}(\nu+1) = \underbrace{[\underline{\Phi}-\underline{H}\ \underline{F}[\underline{I}+\underline{D}\ \underline{F}]^{-1}\underline{C}]}_{\underline{\Phi}_R}\underline{x}(\nu) + [\underline{H}+[\underline{I}+\underline{D}\ \underline{F}]^{-1}\underline{D}]\tilde{\underline{r}}(\nu)$$

$$\underline{y}(\nu) = [\underline{I}+\underline{D}\ \underline{F}]^{-1}\underline{C}\ \underline{x}(\nu) + [\underline{I}+\underline{D}\ \underline{F}]^{-1}\underline{D}\ \tilde{\underline{r}}(\nu) \quad .$$

For the characteristic equation of the closed loop system one obtains (using $\det[\underline{I}+\underline{A}\ \underline{B}] = \det[\underline{I}+\underline{B}\ \underline{A}]$)

$$\det[\underline{I}_n z - \underline{\Phi}+\underline{H}\ \underline{F}[\underline{I}+\underline{D}\ \underline{F}]^{-1}\underline{C}] =$$

$$= \det[\underline{I}_n z-\underline{\Phi}]\det[\underline{I}+[\underline{I}_n z-\underline{\Phi}]^{-1}\underline{H}\ \underline{F}[\underline{I}+\underline{D}\ \underline{F}]^{-1}\ \underline{C}]$$

$$= \det[\underline{I}_n z-\underline{\Phi}]\det[\underline{I}+[\underline{I}+\underline{D}\ \underline{F}]^{-1}\underline{C}[\underline{I}_n z-\underline{\Phi}]^{-1}\underline{H}\ \underline{F}]$$

$$= \det[\underline{I}_n z-\underline{\Phi}]\det[\underline{I}+\underline{D}\ \underline{F}]^{-1}\det[\underline{I}+\underline{D}\ \underline{F}+\underline{C}[\underline{I}_n z-\underline{\Phi}]^{-1}\underline{H}\ \underline{F}]$$

$$= \det[\underline{I}_n z-\underline{\Phi}]\det[\underline{I}+\underline{D}\ \underline{F}]^{-1}\det[\underline{I} + \underline{G}_K(z)\underline{F}] \ . \qquad \textbf{(1.41)}$$

Futhermore, one gets

$$\det[\underline{I}+\underline{G}_k(\infty)\underline{F}] = \det[\underline{I}+\underline{D}\ \underline{F}]$$

and from eq.(1.40)

$$\det[\underline{T}_R(z)] = \det[\underline{I}+\underline{G}_K(z)\underline{F}]^{-1}\det[\underline{G}_K(z)] \ .$$

Now it holds (see eq.(1.41))

$$\det[\underline{I}+\underline{G}_K(z)\underline{F}] = \frac{\det[\underline{G}_K(z)]}{\det[\underline{T}_R(z)]}$$

$$= \det[\underline{I}+\underline{G}_K(\infty)\underline{F}]\ \frac{\det[\underline{I}z-\underline{\Phi}_R]}{\det[\underline{I}z-\underline{\Phi}]} \ . \qquad \textbf{(1.42)}$$

These are the representation and relations of the return difference matrix.
There are similar considerations from Cruz et.al.[1], Desoer et.al.[3], Doyle et.al.[6] and Sain et.al.[15].

1.3 STABILITY OF MULTIVARIABLE FEEDBACK SYSTEMS

We do not want to investigate all aspect of stability. Only for time-invariant discrete control system the principal theorems are given here.
Let $\underline{Z}_r(z)\underline{N}_r^{-1}(z)$ be a right coprime factorization of $\underline{G}_K(z)$ and

$\underline{F}(z) = \underline{I}$ (see fig.1.2), then the feedback transfer matrix (reference behaviour) is given by

$$\underline{T}_R(z) = \underline{G}_K(z)[\underline{I}_p+\underline{G}_K(z)]^{-1} = \underline{Z}_r(z)[\underline{N}_r(z)+\underline{Z}_r(z)]^{-1}. \quad \textbf{(1.43)}$$

Since $\underline{Z}_r$ and $[\underline{N}_r+\underline{Z}_r]$ are right coprime (see eq.(1.3)), $\det[\underline{N}_r(z)+\underline{Z}_r(z)]$ is the closed-loop characteristic polynomial.

It is known, see Desoer et.al.[3], that the feedback system is exponentially stable, if and only if

(a) $\underline{G}_K(z)$ is proper and $\det[\underline{I} + \underline{G}_K(\infty)] \neq 0$

and

(b) the polynomial $\det[\underline{N}_r(z)+\underline{Z}_r(z)]$ has all of its zeros in the open unit disk.

From eq.'s (1.42) and (1.43) it follows

$$\det[\underline{I}+\underline{G}_K(z)] = \frac{\det[\underline{N}_r(z)+\underline{Z}_r(z)]}{\det[\underline{N}_r(z)]} . \quad \textbf{(1.44)}$$

Now one finds,

(a) each zero of $\det[\underline{I}+\underline{G}_K]$ is a pole of $\underline{T}_R(z)$,

(b) each pole of $\underline{T}_R(z)$ is either a zero of $\det[\underline{I}+\underline{G}_K]$ or a pole of $\underline{G}_K(z)$

and

(c) all poles of $\underline{T}_R(z)$ are zeros of the polynomial $\det[\underline{N}_r(z)+\underline{Z}_r(z)]$.

Theorem 1.7: Let the open-loop matrix $\underline{G}_K(z)$ proper and $\det[\underline{I}+\underline{G}_K(\infty)] \neq 0$. Then the closed-loop system is exponentially stable if and only if

$$\inf_{|z|\geq 1} |\det[\underline{I}+\underline{G}_K(z)]| > 0 . \quad \textbf{(1.45)}$$

In order to check the condition (1.45) we introduce the angle function

$$\eta(\alpha) := \mathrm{Im}\left[\log\det[\underline{I}+\underline{G}_K(e^{j\alpha})]\right]$$

with $\eta(0) = 0(\pi)$ if $\det[\underline{I}+\underline{G}_K(1)] > 0$ (< 0).

The condition (1.45) is satisfied if and only if

(a) $\det[\underline{I} + \underline{G}_K(\infty)] \neq 0$

(b) $\inf_{\alpha\varepsilon[0,2\pi]} |\det[\underline{I} + \underline{G}_K(e^{j\alpha})]| > 0$

(if $\underline{G}_K(z)$ has a pole on the unit circle then the contour $[0,2\pi]$ must be indented on the inside)

(c) $\eta(2\pi) - \eta(0) = 2\pi\, n_r$
where n_r is the number of poles of $\underline{G}_K(z)$ outside the closed unit disk.

In continuous case one can formulate the stability criterion in the same way if it will change z to s and the image of the unit circle to the image of the jω axis.

If $\lambda_i(\alpha)$ are eigenlocuses of $\underline{G}_K(e^{j\alpha})$ for $i = 1,\ldots,p$, then it holds

$$\det[\underline{I}+\underline{G}_K(e^{j\alpha})] = \prod_{i=1}^{p} [1 + \lambda_i(\alpha)] \qquad \textbf{(1.46)}$$

and

$$\eta(\alpha) = \mathrm{Im}\left[\sum_{i=1}^{p} \log[1 + \lambda_i(\alpha)]\right]. \qquad (1.47)$$

The $\lambda_i(\alpha)$ are called the <u>principal gains</u> of the transfer matrix $\underline{G}_K(z)$ and can be used in Theorem 1.7, if for all $\alpha\varepsilon[0,2\pi]$ each $\lambda_i(\alpha)$ forms a closed path in the complex λ-plane.
Stability criteria for linear time-variant discrete time systems the reader finds an extensive representation and the proofs in [17], G.Ludyk.

1.4 DESIGN OF TIME-DISCRETE MULTIVARIABLE FEEDBACK SYSTEMS

The design problems occur in discrete feedback systems are essentially similar to those existing in the design of time-continuous control systems. In both cases, there is to control a process such that the outputs of the plant will behave according to some presumed performance specifications, compare section 1.3. The application of the different design methods and descriptions (transfer function or state space) of the feedback system is depending on the desired aims, e.g. stabilization, pole placement, insensitive behaviour, robust stability or optimization in the sense of quadratic performance criterion.

It is not our intention to give a complete description and discussion of the design methods. We confine us on the following two methods:

- pole placement design,
- design with quadratic performance criterion.

Pole placement

One important question in the design of controller is the stabilizability problem. Given a state space model

$$\underline{x}(\nu+1) = \underline{\Phi}\ \underline{x}(\nu) + \underline{H}\ \underline{u}(\nu)$$

one has to check whether a linear feedback law

$$u(\nu) = -\ \underline{K}\ \underline{x}(\nu) + \underline{\rho}\ \underline{r}(\nu)$$

makes the closed-loop system stable, i.e. one obtains for the closed-loop system the relation

$$\underline{x}(\nu+1) = [\underline{\Phi} - \underline{H}\ \underline{K}]\ \underline{x}(\nu) + \underline{H}\ \underline{\rho}\ \underline{r}(\nu) \qquad \textbf{(1.48)}$$

and do all the eigenvalues of $[\underline{\Phi} - \underline{H}\ \underline{K}]$ take a negative real parts ? If so, one calls the pair $(\underline{\Phi}, \underline{H})$ stabilizable.

If one locates the eigenvalues of the closed-loop system (1.48) wherever we like (but for complexe eigenvalues appearing in conjugate pairs) by suitable choice of the state regulator then it will be called the pole placement problem.

Theorem 1.8: The eigenvalues of the closed loop system (1.48) may be arbitrarily located if and only if the system $(\underline{\Phi}, \underline{H})$ is completely controllable. If $rk[\underline{W}] = m<n$ (see eq.(1.8)) then (n-m) eigenvalues of $\underline{\Phi}$ are not changeable in the matrix $[\underline{\Phi} - \underline{H}\ \underline{K}]$, regardless of the choice of $\underline{K}$, but the remaining m eigenvalues of $[\underline{\Phi} - \underline{H}\ \underline{K}]$ can be arbitrarily placed.

Proof: see T.Kailath [8].

In general, all n state variables can not be always measured and then the state controller $\underline{K}$ is not applicable without an observer. Therefore we make the realistic assumption that $p<n$ independent linear combinations of the state variables can be measured such that the output of the plant may be represented as

$$y(\nu) = \underline{C}\ x(\nu)\ .$$

Now the question is, what is the simplest controller to result a stable closed loop system ?

Theorem 1.9: Given the completely controllable and observable state space model

$$\underline{x}(\nu+1) = \underline{\Phi}\ \underline{x}(\nu) + \underline{H}\ \underline{u}(\nu)$$

$$\underline{y}(\nu) = \underline{C}\ \underline{x}(\nu)$$

and the output feedback law

$$\underline{u}(\nu) = -\underline{K}_y \underline{y}(\nu) + \underline{\rho}\ \underline{r}(\nu)\ .$$

Let $l := \min[n; (p+r-1)]$ then $l \leq n$ eigenvalues of the system matrix $[\underline{\Phi} - \underline{H}\ \underline{K}_y \underline{C}]$ can be located arbitrarily by suitable choice of $\underline{K}_y$.

The proof can be derived by using of eq.(1.42) with $\underline{D} = \underline{0}$ and $\underline{F} = \underline{I}$.

In the sense of theorem 1.9 the plant $(\underline{\Phi}, \underline{H}, \underline{C})$ is stabilizable if the unchangeable (n-l) eigenvalues of $\underline{\Phi}$ lie inside of the unit circle.

If l is small compared with n it becomes necessary to construct a dynamic controller. In addition to Theorem 1.8 then one needs an observer if p<n output signals are only available. In chapter 2. an observer will be designed and the aim is not only to estimate the n states (or n linear independent combinations of the states) but also to use free parameters of the observer getting a better dynamical behaviour. A separate design of the state controller $\underline{K}$ and observer is possible for nominal state space models of the plant. Here we give only a procedure for the computation of $\underline{K}$. The reader finds the construction of the observer in chapter 2..

Determination of $\underline{K}$

In Theorem 1.5 it was derived a state controller $\tilde{\underline{K}}$, eq.(1.21), for the canonical representation $(\tilde{\underline{\Phi}},\tilde{\underline{H}})$, eq.(1.20), that makes the feedback system time-optimal (without limit for the input $\underline{u}$). The transformations $\underline{T}_c$ and $\underline{Q}$ map the canonical form $(\tilde{\underline{\Phi}}, \tilde{\underline{H}}, \tilde{\underline{K}})$ in the original system $(\underline{\Phi}, \underline{H}, \underline{K}^*)$ in the following manner:

$$\underline{\Phi}_o = [\tilde{\underline{\Phi}}-\tilde{\underline{H}}\ \tilde{\underline{K}}] = \underline{T}_c[\underline{\Phi}-\underline{H}\ \underline{Q}\ \tilde{\underline{K}}\ \underline{T}_c]\underline{T}_c^{-1} = \underline{T}_c[\underline{\Phi}-\underline{H}\ \underline{K}^*]\underline{T}_c^{-1} \quad .$$

From this relation we conclude for the state controller $\underline{K}^*$ in the original system

$$\underline{K}^* = \underline{Q}\ \tilde{\underline{K}}\ \underline{T}_c \quad . \tag{1.49}$$

Now it follows from the eq.'s (1.20) and (1.21):

$$\tilde{\underline{K}} = \begin{bmatrix} \tilde{\underline{k}}_1' \\ \cdot \\ \cdot \\ \cdot \\ \tilde{\underline{k}}_r' \end{bmatrix} = \begin{bmatrix} \underline{t}_1'\underline{\Phi}^{\mu_1}\underline{T}_c^{-1} \\ \cdot \\ \cdot \\ \cdot \\ \underline{t}_r'\underline{\Phi}^{\mu_r}\underline{T}_c^{-1} \end{bmatrix} = \begin{bmatrix} \underline{t}_1'\underline{\Phi}^{\mu_1} \\ \cdot \\ \cdot \\ \cdot \\ \underline{t}_r'\underline{\Phi}^{\mu_r} \end{bmatrix} \underline{T}_c^{-1}$$

and from eq.(1.19)

$$\underline{t}_i' = \underline{e}_i\underline{W}_o^{-1} \qquad i = 1\ldots r \quad .$$

In the system $(\underline{\Phi},\underline{H})$ one gets the time-optimal state controller

$$\underline{K}^* = \underline{Q} \begin{bmatrix} \underline{e}_1' \underline{W}_o^{-1} \underline{\Phi}^{\mu_1} \\ \vdots \\ \underline{e}_r' \underline{W}_o^{-1} \underline{\Phi}^{\mu_r} \end{bmatrix} . \qquad (1.50)$$

All eigenvalues of the feedback system matrix $[\underline{\Phi}-\underline{H}\ \underline{K}^*]$ lie in the origin of the z-plane. A system matrix $[\underline{\Phi}-\underline{H}\ \underline{K}]$ can be determined with other located eigenvalues in the following way:

- replace $\underline{K} = \underline{K}^* + \underline{K}_1$
- $\underline{K}_1$ shift the eigenvalues of $[\underline{\Phi}-\underline{H}\ \underline{K}^*]$ and one gets the relation from

$$\underline{x}(\nu+1) = \underline{\Phi}\ \underline{x}(\nu) - \underline{H}[\underline{K}^* + \underline{K}_1]\ \underline{x}(\nu)$$

$$= [\underline{\Phi}-\underline{H}\ \underline{K}^*]\underline{x}(\nu) - \underline{H}\ \underline{K}_1\underline{x}(\nu)$$

$$= \underline{T}_c^{-1}\ \underline{\Phi}_o\underline{T}_c\underline{x}(\nu) - \underline{T}_c^{-1}\tilde{\underline{H}}\ \underline{Q}^{-1}\ \underline{K}_1\underline{T}_c^{-1}\ \underline{T}_c\underline{x}(\nu)$$

$$\underline{x}(\nu+1) = \underline{T}_c^{-1}\left[\ \underline{\Phi}_o - \tilde{\underline{H}}\ \underbrace{\underline{Q}^{-1}\underline{K}_1\underline{T}_c^{-1}}_{=\ \underline{K}_o}\right]\underline{T}_c\underline{x}(\nu)$$

From the last equation on finds

$$\underline{K}_1 = \underline{Q}\ \underline{K}_o\ \underline{T}_c . \qquad (1.51)$$

If we choose

$$\underline{K}_o = \begin{bmatrix} \underline{k}_{o1}' & & \underline{O}' \\ & \ddots & \\ \underline{O} & & \underline{O} \\ & \ddots & \\ \underline{O}' & & \underline{k}_{or}' \end{bmatrix} \qquad (1.52)$$

with $\underline{k}_{oi} \in \mathbb{R}^{\mu_i}$ then the characteristic polynom of each subsystem can be separately determined by $\underline{k}_{oi}'$, see page 14 $\tilde{\underline{H}}\ \tilde{\underline{K}}$. When the eigenvalues of the feedback system are given the parameter $\underline{k}_{oi}$ can be computed from the equation

$$\det[\underline{I}\, z - \underline{\Phi} + \underline{H}\,\underline{K}] = \det[\underline{I}\, z - \underline{\Phi}_o + \tilde{\underline{H}}\,\underline{K}_o]$$

$$= \prod_{i=1}^{r} [z^{\mu_i} + \underline{k}_{oi}^{'}\underline{z}_{\mu_i}] \qquad (1.53)$$

with $\underline{z}_{\mu_i}^{'} = [z^{\mu_i - 1}, \ldots, z, 1]$.

The equations (1.50) till (1.53) determine a state controller $\underline{K}$ that only the eigenvalues of the feedback system can be placed arbitrarily, not the zeros.

Time-invariant linear-quadratic problem

Here we are only concerned with the design of state controller and the construction of an observer is the same as it will be treating in chapter 2.. Our aim it is to find an optimal linear state controller $u(\nu) = -\underline{K}\, x(\nu)$ for the state space model

$$\underline{x}(\nu+1) = \underline{\Phi}\,\underline{x}(\nu) + \underline{H}\,\underline{u}(\nu) \qquad \underline{x}_o \neq \underline{0}$$

such that the quadratic performance index

$$J[\{u(\nu)\}] = \frac{1}{2} \sum_{\nu=o}^{\infty} \left[\underline{x}^{'}(\nu)\,\underline{\Sigma}\,\underline{x}(\nu) + \underline{u}^{'}(\nu)\,\underline{R}\,\underline{u}(\nu)\right] \quad (1.54)$$

is minimized with $\underline{\Sigma}$ positive semidefinite and $\underline{R}$ positiv definite.

One important requirement of this infinite-time linear regulator design is that the optimal feedback system should be asymptotically stable, i.e. $\lim_{\nu\to\infty} \underline{x}(\nu) = \underline{0}$.

Assume, that the following conditions hold:

- The matrices $(\underline{\Phi},\underline{H})$ are completely controllable or stabilizable by state controller,
- the matrices $(\underline{\Phi},\tilde{\underline{C}})$ are completely observable, where $\tilde{\underline{C}}$ is any matrix such that $\tilde{\underline{C}}\,\tilde{\underline{C}}^{'} = \underline{\Sigma}$, (1.55)

then there are different methods as calculus of variations, Hamiltonian procedure or dynamic programming to determine the linear optimal state control law. Here we formulate a concise representation of the results for this problem without proof. The reader can find the proof in each book about optimal control.

Theorem 1.10: Assume that for a given linear time invariant state space model

$$\underline{x}(\nu+1) = \underline{\Phi}\,\underline{x}(\nu) + \underline{H}\,\underline{u}(\nu) \qquad \underline{x}(0) \neq \underline{0}$$

the conditions (1.55) are fulfilled. Then the control law

$$\underline{u}^*(\nu) = -[\underline{R} + \underline{H}'\underline{P}\,\underline{H}]^{-1}\,\underline{H}'\underline{P}\,\underline{\Phi}\,\underline{x}(\nu) \qquad (1.56)$$

$$= -\underline{K}^*\,\underline{x}(\nu)$$

is optimal in the sense of the quadratic performance index (1.54) and the equilibrium point $\underline{x}_e = \underline{0}$ of the feedback system is asymptotically stable. Furthermore $\underline{P} = \underline{P}'$ is a positiv definite unique solution of the steady state (algebraic) Riccati equation

$$\underline{P} = \underline{\Sigma} + \underline{\Phi}'\underline{P}\,\underline{\Phi} - [\underline{\Phi}'\underline{P}\,\underline{H}][\underline{R} + \underline{H}'\underline{P}\,\underline{H}]^{-1}\,\underline{H}'\underline{P}\,\underline{\Phi}$$

$$= \underline{\Sigma} + \underline{\Phi}'\underline{P}[\underline{\Phi} - \underline{H}\,\underline{K}^*] \qquad (1.57)$$

and the minimum of the performance index is given by

$$J^*[\underline{x}(0)] = \frac{1}{2}\,\underline{x}'(0)\underline{P}\,\underline{x}(0) \quad . \qquad (1.58)$$

The solution of the Riccati equation (1.57) can be solved recursively starting with $\underline{P}_o = \underline{0}$, i.e.

$$\underline{P}_{\mu+1} = \underline{\Sigma} + \underline{\Phi}'\underline{P}_\mu\underline{\Phi} - [\underline{\Phi}'\underline{P}_\mu\underline{H}][\underline{R} + \underline{H}'\underline{P}_\mu\underline{H}]^{-1}\underline{H}'\underline{P}_\mu\underline{\Phi} \; .$$

$$\mu = 0, 1, 2, \ldots$$

The generated sequence $\{\underline{P}_\mu\}$ converges under the conditions (1.55) and it holds

$$\underline{P} = \lim_{\mu\to\infty} \underline{P}_\mu \quad .$$

For a given small value ε the sequence will be truncated such that for a suitable norm one uses the unequation

$$||\underline{P}_{\mu+1} - \underline{P}_\mu|| < \varepsilon \; .$$

It should be noted that truncation (depending from ε) and roundoff errors appear in the computation of $\underline{P}$.

Sensitivity behaviour

The Riccati equation (1.57) can be transformed in a relation of z-transfer functions after performing in some steps. With the open loop transfer matrix

$$\underline{L}_r(z) = \underline{K}^*[\underline{I}_n z - \underline{\Phi}]^{-1}\,\underline{H}\quad,\quad \det[\underline{I}_r+\underline{L}_r]\text{see eq.(1.42)},$$

$\underline{\Sigma} = \tilde{C}\,\underline{\tilde{C}}'$ and

$$\underline{\tilde{G}}(z) = \underline{\tilde{C}}'[\underline{I}_n z - \underline{\Phi}]^{-1}\,\underline{H}$$

one obtains

$$\left[\underline{I}_r+\underline{L}_r(z^{-1})\right]'\,\underline{R}\left[\underline{I}_r+\underline{L}_r(z)\right] = \underline{R} + \underline{\tilde{G}}'(z^{-1})\underline{\tilde{G}}(z). \qquad (1.59)$$

The reader should verify this relation. If we determine the singular values of eq.(1.59) in the frequency domain one gets with $\underline{R} = \underline{I}_r\gamma$

$$\sigma_i[\underline{I}_r+\underline{L}_r(e^{j\varphi})] = \sqrt{\lambda_i\left[\underline{I}_r+\frac{1}{\gamma}\,\underline{\tilde{G}}'(e^{-j\varphi})\underline{G}(e^{j\varphi})\right]}$$

$$= \sqrt{1+\frac{1}{\gamma}\,\sigma_i^2[\underline{\tilde{G}}]}\quad. \qquad (1.60)$$

An estimation of the return difference matrix lead us to the unequation, see also eq.(1.35),

$$\underline{\sigma}[\underline{I}_r+\underline{L}_r] = \frac{1}{\bar{\sigma}[\underline{S}]} > 1 \qquad (1.61)$$

for all frequencies. If it holds $\frac{1}{\sqrt{\gamma}}\,\underline{\sigma}[\underline{\tilde{G}}] >> 1$, then one obtains from eq.(1.60) the estimation

$$\underline{\sigma}[\underline{L}_r] \approx \frac{1}{\sqrt{\gamma}}\,\underline{\sigma}[\underline{\tilde{G}}] \qquad (1.62)$$

Hence, the optimal state feedback system are guaranteed to remain stable for uncertainties $\bar{\sigma}[\Delta\underline{L}] < 1$, if it will be chosen suitable γ and $\underline{\tilde{C}}$, see also eq.(1.38).

1.5 REFERENCES

[1] J.B.Cruz, J.S.Freudenberg and D.P.Looze:"A relationship between sensitivity and stability of multivariable feedback systems", JACC 1980-WP8-C.

[2] P.Brunowsky:"A classification of linear controllable systems", Kybernetica,Cislo,pp.173-188,1970.

[3] C.A.Desoer and M.Vidyasagar:"Feedback systems; input-output properties", Academic Press New York 1975.

[4] C.A.Desoer and Y.T.Wang:"On the generalized Nyquist stability criterion", JACC 1979-TP5-2:30, pp.580-586.

[5] A.Dickman and R.Sivan:"On the robustness of multivariable linear feedback systems", IEEE Trans.Automat.Contr.,Vol.AC-30,No.4,pp.401-404,1985.

[6] J.C.Doyle and G.Stein:"Multivariable feedback design: Concepts for a classical/modern synthesis", IEEE Trans.Automat.Contr.,Vol.AC-26,No.1, pp.4-16, 1981.

[7] B.A.Francis:"On robustness of the stability feedback systems", IEEE Trans.Automat.Contr.,Vol.AC-25, No.4, pp.817-818, 1980.

[8] T.Kailath:"Linear systems", Prentice-Hall, Englewood Cliffs, N.J., 1980.

[9] H.W.Knobloch and H.Kwakernaak:"Lineare Kontrolltheorie" Springer-Verlag Berlin 1985.

[10] A.J.Laub:"An inequality and some computations related to robust stability of linear dynamic systems", IEEE Trans.Automat.Contr.,Vol.AC-24, No.2, pp. 318-320.

[11] D.W.Nuzman and N.R.Sandell,JR.:"An inequality in robustness analysis of multivariable systems", IEEE Trans.Automat.Contr.,Vol.AC-24,No.3, pp.492-493, 1979.

[12] D.H.Owens and A.Raya:"Robust stability of Smith predictor controllers for time-delay systems", IEE Proc., Vol.129, No.6, pp.298-304, 1982.

[13] D.H.Owens and A.Chotai:"Robust stability of multivariable feedback systems with respect to linear and nonlinear feedback perturbations", IEEE Trans. Automat.Contr.,Vol.AC-27, No.1,pp.254-256,1982.

[14] V.M.Popov:"Invariant description of linear time-invariant controllable systems", SIAM J.Control, Vol.10,pp.252-264, 1972.

[15] M.K.Sain and A.Ma:"Multivariable synthesis with reduced comparison sensitivity", JACC 1980-WP8-B.

[16] G.Zames:"Feedback and optimal sensitivity; model reference transformations, multiplicative seminorms and approximate inverses", IEEE Trans. Automat.Contr.,Vol.AC-26,No.2,pp.301-320,1981.

[17] G.Ludyk:"Stability of time-variant discrete-time systems", Vol.5, series: Advances in Control Systems and Signal Processing, Vieweg Braunschweig/ Wiesbaden 1985.

2 Design of Linear Time-Invariant Feedback Systems with Minimized Comparison Sensitivity Function

2.1 INTRODUCTION

The process to be controlled shall be representable by means of a linear discrete-time state space model

$$\begin{aligned} \underline{x}_{\nu+1} &= \underline{\Phi}\,\underline{x}_\nu + \underline{H}\,\underline{u}_\nu \\ \underline{y}_\nu &= \underline{C}\,\underline{x}_\nu + \underline{D}\,\underline{u}_\nu \quad , \end{aligned} \tag{2.1}$$

wherein:

state variable $\underline{x}_\nu \in \mathbb{R}^n$

input variable $\underline{u}_\nu \in \mathbb{R}^r$

output variable $\underline{y}_\nu \in \mathbb{R}^p$.

It is possible to make direct use of a control law of the form

$$\underline{u}_\nu = -\underline{K}\,\underline{x}_\nu + \underline{\rho}\,\underline{r}_\nu \tag{2.2}$$

with

command input $\underline{r}_\nu \in \mathbb{R}^{\tilde{r}}$, $\tilde{r} \leq \min(r,p)$

provided that all state variables are measurable; if not, it will be necessary to make use of a dynamic state regulator. A dynamic state regulator (comprising, for example, a state observer and the gain matrix $\underline{K}$) will furnish the control law (2.2) exactly provided that the plant has been based upon the precise model (2.1) at the time of designing and that no disturbance inputs are able to apply. A distinguishing feature, therefore, of the dynamic state regulator is the fact that the nominal command behavior $\underline{R}(z) \rightarrow \underline{Y}(z)$ is dependent only upon $\underline{K}$ [1] .

The command behavior normally changes in relation to the nominal command behavior established in Eq.(2.2) if the process to be controlled is disturbed, i.e. if (2.1) is changed into the disturbed state space model (cf. Fig.1):

$$\underline{x}_{\nu+1} = \underline{\Phi}\,\underline{x}_\nu + \underline{H}\,\underline{u}_\nu + \underline{E}_1\,\underline{\xi}_\nu$$
$$\underline{y}_\nu = \underline{C}\,\underline{x}_\nu + \underline{D}\,\underline{u}_\nu + \underline{E}_2\,\underline{\xi}_\nu \qquad (2.3)$$

with the
disturbance input $\underline{\xi}_\nu \in \mathbb{R}^\xi$, ξ arbitrary
and the parameter variations

$\underline{\Phi} \to \underline{\Phi} + \Delta\underline{\Phi}$, $\underline{H} \to \underline{H} + \Delta\underline{H}$

$\underline{C} \to \underline{C} + \Delta\underline{C}$, $\underline{D} \to \underline{D} + \Delta\underline{D}$.

It is assumed that the disturbance input $\underline{\xi}_\nu$ satisfies a linear difference equation of finite order, the characteristic values of which are known and do not lie within the unit circle, i.e. disturbance inputs are being considered, of which the influence on $\nu \to \infty$ does not fade away spontaneously. The parameter variations $\Delta\underline{\Phi}$, $\Delta\underline{H}$, $\Delta\underline{C}$ and $\Delta\underline{D}$ are not regarded as being necessarily infinitesimal. They are, however (as a function of the example for application), limited as to their upper values by the fact that the feedback system requires stability.

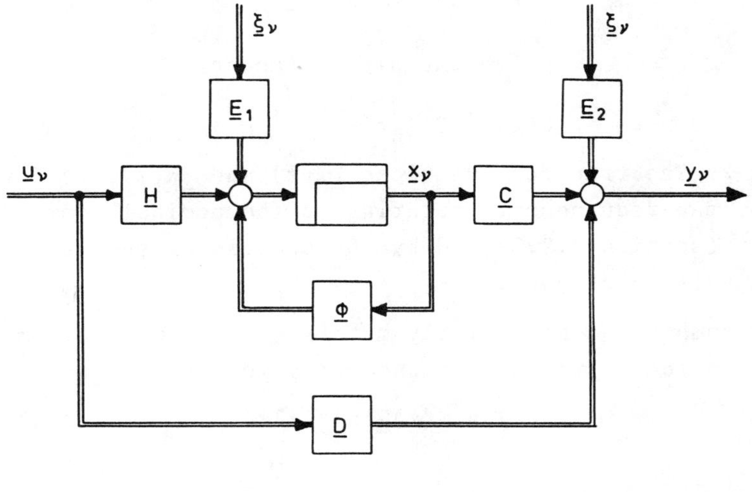

Fig.2.1: Disturbed process

It is now important that the controller be made subject to the minimum requirement that for the command behavior the influence of constant parameter variations (i.e. plant parameters are constant but not precisely known) and the influence of disturbance inputs on the output variable $\underline{y}_\nu$ for $\nu \rightarrow \infty$ should disappear.

The point of departure for solving this problem in the present work is the comparison sensitivity matrix $\underline{S}(z)$. It indicates the factor by which the sensitivity of the closed control loop differs from the sensitivity of an assumed open loop compensator which, in the absence of parameter variations, furnishes the same command behavior as does the closed control loop (nominally equivalent open loop compensation):

$$\underline{E}_R(z) = \underline{S}(z)\underline{E}_S(z) \quad . \tag{2.4}$$

Here, $\underline{E}_R(z)$ and $\underline{E}_S(z)$ are the deviations in the output variable caused by parameter variations in the cases of the closed control loop and the nominally equivalent open loop compensator.
The comparison sensitivity matrix $\underline{S}(z)$ is provided by the inverse "effect of control" [2]:

$$\underline{S}(z) = \left\{\underline{I}_p + \underline{G}(z)\underline{\Gamma}_C^{-1}(z)\underline{Z}_C(z)\right\}^{-1} \tag{2.5}$$

with the plant transfer function

$$\underline{G}(z) = \underline{C}(\underline{I}_n z - \underline{\Phi})^{-1}\,\underline{H} + \underline{D}$$

cf. Fig.2. The transfer functions $\underline{\Gamma}_C(z)$ and $\underline{Z}_C(z)$ are determined by the requirements relating to the nominal command behavior (section 2.2.3) and by the additional sensitivity requirements (section 2.3).

The comparison sensitivity matrix can likewise be employed for evaluation of the disturbance behavior $\underline{\Xi}(z) \rightarrow \underline{Y}(z)$:

$$\underline{Y}(z) = \underline{S}(z)\hat{\underline{G}}(z)\underline{\Xi}(z) \quad . \tag{2.6}$$

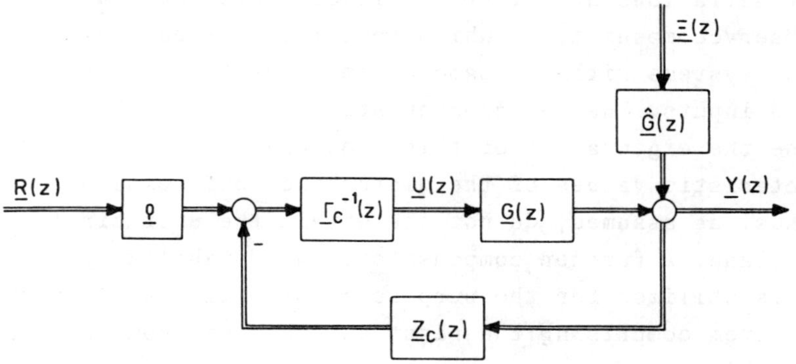

Fig.2.2: Structure of the disturbed feedback system in the frequency domain (section 2.2.7.2)

Here,

$$\underline{\hat{G}}(z) = \underline{C}(\underline{I}_n z - \underline{\Phi})^{-1}\underline{E}_1 + \underline{E}_2$$

represents the transfer function of the disturbance input $\underline{\xi}_\nu$ to $\underline{y}_\nu$ in Fig.1. The pole positions of $\underline{\hat{G}}(z)$ cancel themselves out in Eq.(2.6) against corresponding zeros of $\underline{S}(z)$.

Sections 2.4 and 2.5 deal with the development of a process enabling, first of all in accordance with section 2.5.1 the generation of zeros (of arbitrary multiplicity) for the sensitivity function $\underline{S}(z)$ (provided that they are different from the zeros of the plant transfer function $\underline{G}(z)$, cf. section 2.2.7.2.3). Should these zeros be so selected as to be equal to the characteristic values of the disturbance input and of the command input, then, in accordance with Eqs.(2.4) and (2.6), the above-mentioned minimum requirement will be satisfied. (It should be noted with regard to Eq.(2.4) that $\underline{E}_S(z)$ possesses the same characteristic values as the command input).

The minimum requirement in question has been dealt with on various occasions in literature.

Davison, in a series of works concluded with [3], deals with the extension of the classical output servo control principle to multivariable systems, whereby the $\tilde{r}$ ($\tilde{r} \leq \min(r,p)$)

output deviations are first of all conducted over a compensator ("servocompensator") which comprises $\tilde{r}$ identical linear partial systems with the same dynamics as the disturbance and command inputs. The servocompensator is an unstable system because the eigenvalues of this compensator coincide with the characteristic values of the disturbance and command inputs and thus, as assumed, do not lie within the unit circle of the z-plane. A further compensator, the "stabilizing compensator", is utilized for the purpose of stabilizing the overriding system comprising the plant and the servocompensator. This stabilizing compensator possesses as input variables the outputs of the plant and servocompensator and its output variable forms the control input of the plant.

The servocompensator directly enables the sensitivity function $\underline{S}(z)$ to possess zeros at the points occupied by the characteristic values of the disturbance and command inputs. As the dynamics of the command input are also incorporated into the sensitivity design, the control variables for $t\to\infty$ indeed act against the reference value. This, however, does not result in a determining of the nominal command behavior separate from the sensitivity design and the resulting feedback system behavior will often be unsatisfactory.

Young and Willems [4] likewise employ a servocompensator of the type described, though they limit themselves to the special case of constant-time disturbance inputs.

Noldus [5] gives further treatment to the problem of disturbance rejection for the whole plant state vector. Here, with the exception of special cases, disturbance inputs are only permissible where these apply at the plant input. A compensator is designed in the frequency domain in such a manner as to enable the zeros of the disturbance transfer function (i.e. the transfer function of the input disturbance variable to the plant state vector) to correspond to the characteristic values of the disturbance input.
Even this procedure does not allow the possibility of determining the nominal feedback system behavior independently of

the sensitivity requirements. The author states that it is intended to generalize the design so as to extend it to multivariable systems.

A more frequently used design principle involves the procedure of disturbance rejection including estimation and counteraction. Here, the plant and disturbance input models are combined to provide an extended state space model and a state observer is employed for reconstructing the plant state and the disturbance input.
The reconstructed disturbance input is then used for the disturbance rejection whilst the observed plant state is able to be utilized for realizing a control law of form (2.2). This design principle was first mentioned by Johnson [eg. 6] .

Research by Mikolcic [7], in which an example was used to make a comparison of the results obtained through the methods of procedure employed by Davison and Johnson, has unfailingly provided a more favorable feedback system behavior when using Johnson's procedure, both where disturbance inputs are present and with regard to parameter variations. The reason for this will be the fact that with Johnson the nominal feedback system behavior is (with the aid of the state gain matrix $\underline{K}$) determined independently of the sensitivity behavior.

The most generalized treatment of the procedure of disturbance rejection including estimation and counteraction is given by Müller and Lückel [8]. They enlarge upon Johnson by also considering plants with a distribution matrix and, furthermore, they permit the application of disturbance inputs at the plant output and introduce a (disturbed) control variable vector.

In all procedures which permit disturbance inputs at the plant output it is basically possible to apply the (negative) command input at the plant output and, for the sensitivity design, to theoretically add the command input to the disturbance inputs applying at the plant output (cf.section 2.2.7.5). In all cases, the command transfer function $\underline{I}_p$ - $\underline{S}(z)$ will then follow, from which it becomes evident that it is not possible to separately determine the sensitivity behavior and the command behavior. In normal cases this will result in an unfavorable command behavior.

The procedure of disturbance rejection including estimation and counteraction also envisages the possibility of the number of control inputs falling below the number of outputs in question. Then, however (with the exception of special cases), disturbance inputs are only permissible where these apply at the plant input and nor do the control variables conducted show any parameter insensitivity.

A further method of solution, which is based on geometrical considerations, is indicated by Sebakhy and Wonham [9]. Here, plants without a distribution matrix are considered.

The method of procedure used in the present work is related to the principle of disturbance rejection including estimation and counteraction with additional consideration of the parameter insensitivity of the output variables conducted. During the design, however, a state space model of the disturbance is not added explicitly to the state space model of the plant but, instead, the controller is designed in accordance with sensitivity behavior requirements which have been formulated in the frequency domain.

A special characteristic of this work is the direct orientation of the design to the comparison sensitivity function $\underline{S}(z)$. This makes it possible, among other things, to satisfy the minimum requirement already dealt with in literature (zeros for $\underline{S}$) and, moreover, to minimize the norm of the sensitivity matrix $\underline{S}(z)$ within a desired frequency range (section 2.5.2). This is of significance for a number of reasons:

* Not only are disturbance inputs permitted which, as required at the outset, represent solution functions of a linear difference equation but, in addition, there is a reduction in the influence of disturbance

inputs of which the spectrum extends over a continuous frequency range (in which $||\underline{S}||$ will have been minimized).

* The minimization of the sensitivity function norm in the lower to middle frequency range generally results in an improved transient response of the feedback system.

The free parameters in the controller necessary for the minimizing and/or generation of the zeros of $\underline{S}(z)$ are obtained through a corresponding increasing of the order of the controller. This is so because the quantity of free parameters increases with the controller order (section 2.5.). The effect of the minimization depends on the quantity of free parameters made available in this manner. The greater the number of parameters available for the minimization, the smaller the values assumed by the norm of the sensitivity matrix $\underline{S}(z)$ in the relevant frequency range. It should be noted that in normal cases this will result in an increase in the values within the remaining frequency range, whereby (depending on the particular application) the controller order may be limited as to its upper level.

2.2 LINEAR MULTIVARIABLE FEEDBACK SYSTEMS WITH DYNAMIC STATE REGULATOR

Following the description of the feedback system structure (section 2.2.1) and of the general requirements to be made regarding the behavior of the feedback system (section 2.2.2) sections 2.2.3 to 2.2.5 shall deal with the controller design with regard to provision of the control law (2.2) applicable in nominal cases. Section 2.2.6 will then discuss the free parameters which thereafter remain in the controller (these being as yet undetermined parameters which have no influence upon the nominal command behavior) and their quantity. Finally, section 2.2.7 has as its theme the structure of the feedback system within the frequency domain and provides the basis for the more detailed specification, to be dealt with in the third section, of the sensitivity requirements and their translation into requirements regarding controller transfer functions.

2.2.1 STRUCTURE OF THE FEEDBACK SYSTEM

The subsequently utilized structure of the feedback system is shown in Fig.3. The controller comprises r dynamic partial controllers, each of which possesses the following state-regulator observer structure:

$$\begin{aligned} \underline{v}_{\nu+1}^{(i)} &= \underline{F}^{(i)} \underline{v}_{\nu}^{(i)} + \underline{S}_{y}^{(i)} (\underline{y}_{\nu} - \underline{D}\, \underline{u}_{\nu}) + \underline{S}_{u}^{(i)} \underline{u}_{\nu} \\ w_{\nu}^{(i)} &= \underline{k}_{m}^{'(i)} \underline{v}_{\nu}^{(i)} + \underline{k}_{y}^{'(i)} (\underline{y}_{\nu} - \underline{D}\, \underline{u}_{\nu}) \quad , \end{aligned} \tag{2.7}$$

$\underline{v}_{\nu}^{(i)} \in \mathbb{R}^{m}$, $i = 1,\ldots r$. The following will then be applicable for the control input $\underline{u}_{\nu}$:

$$\underline{u}_{\nu} = - \underline{w}_{\nu} + \underline{\rho}\, \underline{r}_{\nu} \tag{2.8}$$

$$\text{with} \quad \underline{w}_{\nu} = \begin{bmatrix} w_{\nu}^{(1)} \\ w_{\nu}^{(2)} \\ \cdot \\ \cdot \\ \cdot \\ w_{\nu}^{(r)} \end{bmatrix} .$$

It can be seen from Fig.3 that under nominal command behavior (no disturbance inputs or parameter variations) the influence of the distribution matrix $\underline{D}$ on the controller is fully compensated. For this reason, the distribution matrix is left out of consideration in section 2.2.2 to 2.2.5 which have as their theme the nominal command behavior.

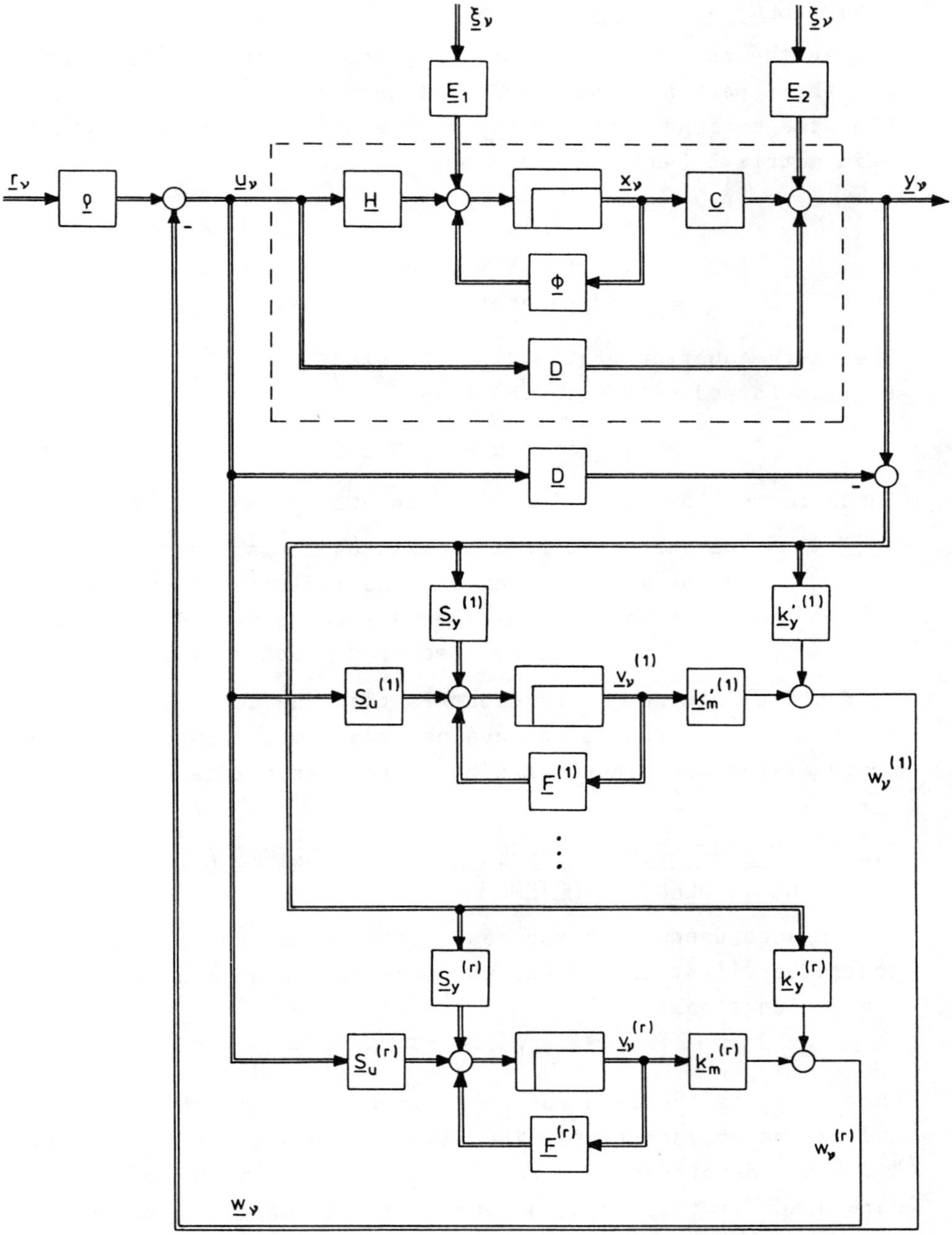

Fig.2.3: Structure of feedback system

2.2.2 GENERAL REQUIREMENTS

In the case of the feedback system comprising the plant and the r partial controllers, it should first of all be true*) that the command behavior ($\underline{x}_o = \underline{0}$) will be determined by the gain matrix $\underline{K}$ (and the prefactor $\underline{\rho}$), i.e.

$$\underline{w}_\nu \equiv \underline{K}\ \underline{x}_\nu \qquad (2.9)$$

in the absence of disturbances and parameter variations.

The state equation of the discrete plant model (2.1) will then be transformed with Eq.(2.8) into

$$\underline{x}_{\nu+1} = (\underline{\Phi} - \underline{H}\ \underline{K})\underline{x}_\nu + \underline{H}\ \underline{\rho}\ \underline{r}_\nu \quad . \qquad (2.10)$$

Furthermore, the following requirements are made in this work:

> The influence of the disturbance inputs defined in the Introduction and the influence of parameter variations in the plant are to be minimized in a manner still to be specified (section 2.3).

A practical minimum requirement is that the control variable $\underline{y}_\nu$ for $\nu \to \infty$ should act against the command input $\underline{r}_\nu$ even where disturbance inputs and parameter variations occur.

2.2.3 INTERRELATIONS BETWEEN PLANT PARAMETERS AND CONTROLLER PARAMETERS

In accordance with the requirement (2.9) the i-th partial controller first of all has the task of making available the linear functional

$$w_\nu^{(i)} \equiv \underline{k}_i'\ \underline{x}_\nu \qquad (2.11)$$

wherein $\underline{k}_i'$ is the i-th row of $\underline{K}$, in the absence of disturbances and parameter variations. The elucidations as far as section 2.2.5 are devoted solely to this aspect. The method of procedure used for this does, however, have a bearing upon the other requirements of section 2.2.2 which are described in greater detail in section 2.3.

*) in accordance with Eq.(2.2)

The following theorem is of significance with regard to the requirement (2.11):

On the assumption that the plant (2.1) is complete controllable and that the system (2.7) is complete observable, then

$$\lim_{\nu\to\infty} w_{\nu+j}^{(i)} = \lim_{\nu\to\infty} \underline{k}_i' \, \underline{x}_{\nu+j} \quad , \quad j = 0,1,2,... \qquad (2.12)$$

will be applicable independently of $\underline{x}_o$, $\underline{v}_o$ and $\{\underline{u}_\nu\}$ if and only if a matrix $\underline{Q}^{(i)}$ existing, so that the following conditions can be satisfied:

(a) $\underline{S}_u^{(i)} = \underline{Q}^{(i)} \underline{H}$

(b) $\underline{Q}^{(i)} \underline{\Phi} - \underline{F}^{(i)}\underline{Q}^{(i)} = \underline{S}_y^{(i)}\underline{C}$

(c) $\underline{k}_m^{'(i)}\underline{Q}^{(i)} + \underline{k}_y^{'(i)}\underline{C} = \underline{k}_i'$ $\qquad$ (i = 1,..,r) (2.13)

(d) all eigenvalues of $\underline{F}^{(i)}$ in the unit circle.

The proof of the above theorem for the continuous-time case is to be found in [11] and can be easily transferred to the discrete-time case.

Subsequently, the complete observability of the plant will also be assumed.

2.2.4 DESIGN RELATIONSHIPS IN A CANONICAL FORM FOR THE PLANT

First of all, the results of a regular state space transformation of the plant model (2.1) shall be shown.

By means of a state space transformation

$$\underline{x}_\nu = \underline{T} \, \tilde{\underline{x}}_\nu \qquad (2.14)$$

the plant model (2.1) will assume the form

$$\begin{aligned} \tilde{\underline{x}}_{\nu+1} &= \tilde{\underline{\Phi}} \, \tilde{\underline{x}}_\nu + \tilde{\underline{H}} \, \underline{u}_\nu \\ \underline{y}_\nu &= \tilde{\underline{C}} \, \tilde{\underline{x}}_\nu + \underline{D} \, \underline{u}_\nu \end{aligned} \qquad (2.15)$$

with the transformed matrices

$$\tilde{\underline{\Phi}} = \underline{T}^{-1}\underline{\Phi}\,\underline{T}$$
$$\tilde{\underline{H}} = \underline{T}^{-1}\underline{H} \qquad (2.16)$$
$$\tilde{\underline{C}} = \underline{C}\,\underline{T} \quad .$$

Therefore, the design relationships (2.13) will be transformed into

$$\text{(a)} \quad \underline{S}_u^{(i)} = \tilde{\underline{Q}}^{(i)}\tilde{\underline{H}}$$
$$\text{(b)} \quad \tilde{\underline{Q}}^{(i)}\tilde{\underline{\Phi}} - \underline{F}^{(i)}\tilde{\underline{Q}}^{(i)} = \underline{S}_y^{(i)}\tilde{\underline{C}} \qquad (i = 1,..r) \quad (2.17)$$
$$\text{(c)} \quad \underline{k}_m^{'(i)}\tilde{\underline{Q}}^{(i)} + \underline{k}_y^{'(i)}\tilde{\underline{C}} = \tilde{\underline{k}}_i^{'} \quad ,$$

wherein $\tilde{\underline{Q}} = \underline{Q}\,\underline{T}$ is applicable and $\tilde{\underline{k}}_i^{'}$ is the i-th row of the transformed gain matrix

$$\tilde{\underline{K}} = \underline{K}\,\underline{T} \quad . \qquad (2.18)$$

Henceforth, the indices (i) shall be omitted for the sake of clarity, i.e.

$$\text{(a)} \quad \underline{S}_u = \tilde{\underline{Q}}\,\tilde{\underline{H}}$$
$$\text{(b)} \quad \tilde{\underline{Q}}\,\tilde{\underline{\Phi}} - \underline{F}\,\tilde{\underline{Q}} = \underline{S}_y\tilde{\underline{C}} \qquad (2.19)$$
$$\text{(c)} \quad \underline{k}_m^{'}\tilde{\underline{Q}} + \underline{k}_y^{'}\tilde{\underline{C}} = \tilde{\underline{k}}^{'} \quad .$$

In the case of the i-th partial controller, $\tilde{\underline{k}}^{'}$ can thus be interpreted as the i-th row of $\tilde{\underline{K}}$ (the sole difference in the design of the individual partial controllers).

The plant should now lie in Luenberger's second form (also termed "observable canonical form") before [12] :

$$\tilde{\underline{\Phi}} = \left[\underline{i}_2 \cdots \underline{i}_{n_1}, \underline{a}_1 \middle| \underline{i}_{n_1+2} \cdots \underline{i}_{n_1+n_2}, \underline{a}_2 \middle| \cdots \underline{i}_n, \underline{a}_p\right] \qquad (2.20)$$

$\tilde{\underline{H}}$ without particular form

$$\tilde{\underline{C}} = \underline{P}\,\hat{\underline{C}} \qquad (2.21)$$

with the triangular matrix

$$\underline{P} = \begin{bmatrix} 1 & & & \underline{0} \\ p_{21} & 1 & & \\ \vdots & & \ddots & \\ p_{p1} & p_{p2} & \cdots & 1 \end{bmatrix} \qquad (2.22)$$

and

$$\hat{\underline{C}} = \begin{bmatrix} \underline{i}'_{n_1} \\ \underline{i}'_{n_1+n_2} \\ \vdots \\ \underline{i}'_{n} \end{bmatrix}. \qquad (2.23)$$

Here, $\underline{i}_\mu$ denotes the μ-th unit vector of the dimension n.

The state regulator row $\tilde{\underline{k}}'$ shall be represented in the form

$$\tilde{\underline{k}}' =: \left[k_1^{(1)}, \ldots, k_{n_1}^{(1)} \middle| \, k_1^{(2)}, \ldots, k_{n_2}^{(2)} \, \middle| \ldots \middle| \, k_1^{(p)}, \ldots, k_{n_p}^{(p)} \right] . \qquad (2.24)$$

If we now insert $\tilde{\underline{\Phi}}$, $\tilde{\underline{C}}$ and $\tilde{\underline{k}}'$ according to (2,20),(2.21) and (2.24) into the design conditions (2.19b) and (2.19c) then we shall obtain the following after an elementary intermediate calculation *) :

$$\tilde{\underline{Q}} = \left[\underline{q}_1, \underline{F}\, \underline{q}_1, \ldots, \underline{F}^{n_1-1} \underline{q}_1 \middle| \ldots \middle| \, \underline{q}_p, \underline{F}\, \underline{q}_p, \ldots, \underline{F}^{n_p-1} \underline{q}_p \right] \qquad (2.25)$$

$$\underline{S}_y = \left[\tilde{\underline{Q}}\, \underline{a}_1 - \underline{F}^{n_1} \underline{q}_1 \middle| \; . \; . \; . \middle| \, \tilde{\underline{Q}}\, \underline{a}_p - \underline{F}^{n_p} \underline{q}_p \right] \underline{P}^{-1} \qquad (2.26)$$

$$\underline{k}'_y = \left[k_{n_1}^{(1)} - k_{n_1}^{*(1)}, \ldots, k_{n_p}^{(p)} - k_{n_p}^{*(p)} \right] \underline{P}^{-1} \qquad (2.27)$$

and

---

*) a more detailed description is given in [15].

$$\begin{bmatrix} \underline{q}_1' \\ \underline{q}_1'\underline{F}' \\ \vdots \\ \underline{q}_1'\underline{F}'^{n_1-2} \\ \hline \vdots \\ \hline \underline{q}_p' \\ \underline{q}_p'\underline{F}' \\ \vdots \\ \underline{q}_p'\underline{F}'^{n_p-2} \end{bmatrix} \underline{k}_m = \begin{bmatrix} k_1^{(1)} \\ k_2^{(1)} \\ \vdots \\ k_{n_1-1}^{(1)} \\ \hline \vdots \\ \hline k_1^{(p)} \\ k_2^{(p)} \\ \vdots \\ k_{n_p-1}^{(p)} \end{bmatrix} . \qquad (2.28)$$

In (2.27) the abbreviation

$$k_{n_i}^{*(i)} := \underline{k}_m' \, \underline{F}^{n_i-1} \underline{q}_i \qquad (2.29)$$

was used.

The problem of dimensioning of the controller in order to satisfy the requirement (2.9) will thus have become simplified to the determining of

$$\underline{F},\ \underline{k}_m' \text{ and the vectors } \underline{q}_1, \ldots \underline{q}_p ,$$

so that the set of equations (2.28) will be satisfied (section 2.2.5).
The controller elements

$$\underline{S}_u,\ \underline{S}_y \text{ and } \underline{k}_y$$

will by then conform fully to (2.25),(2.26),(2.27) and (2.19a).

Up to this point, the design represents a generalization of the method of procedure applied in [11] to the multivariable case.

2.2.5 CANONICAL FORM FOR THE CONTROLLER

The following p sets of equations will be obtained after a row-by-row transposition of Eq. (2.28)

$$\begin{bmatrix} \underline{k}_m' \\ \underline{k}_m'\underline{F} \\ \cdot \\ \cdot \\ \cdot \\ \underline{k}_m'\underline{F}^{n_i-2} \end{bmatrix} \underline{q}_i = \begin{bmatrix} k_1^{(i)} \\ k_2^{(i)} \\ \cdot \\ \cdot \\ \cdot \\ k_{n_i-1}^{(i)} \end{bmatrix}, \quad i = 1,\dots p . \tag{2.30}$$

It is assumed that each partial controller will be complete observable. Therefore, there should be no restriction in the dynamic behavior of a partial controller if the pair $(\underline{F}, \underline{k}_m')$ is arranged in a complete observable canonical form.

With

$$\underline{F} = \begin{bmatrix} 0 & 1 & & & \underline{0} \\ \cdot & & & & \\ \cdot & & & & \\ \cdot & & & & \\ 0 & 0 & \cdot & \cdot & \cdot & 1 \\ -\beta_o & -\beta_1 & \cdot & \cdot & \cdot & -\beta_{m-1} \end{bmatrix} \tag{2.31}$$

$$\underline{k}_m' = \begin{bmatrix} 1 & 0 & \cdot & \cdot & \cdot & 0 \end{bmatrix} \tag{2.32}$$

the following is applicable:

$$\begin{bmatrix} \underline{k}_m' \\ \underline{k}_m'\underline{F} \\ \cdot \\ \cdot \\ \cdot \\ \underline{k}_m'\underline{F}^{m-1} \end{bmatrix} = \underline{I}_m \quad , \tag{2.33}$$

wherein $\underline{I}_m$ is the m-series unit matrix. n_o shall now be the observability index of the discrete plant:

$$n_o := \max_i n_i \quad . \tag{2.34}$$

Henceforth, $m \geq n_o-1$ will always be arranged for the order of any one partial controller, this being because in this case all p sets of equations (2.30) will have a solution, independently of $\beta_o, \dots \beta_{m-1}$, as is shown below.
Either $n_i-1 = m$ or $n_i-1 < m$ can be applied for the structural indices n_i in the individual sets of equations (2.30).

Taking into consideration Eq. (2.33), a set of equations from (2.30) with $n_i-1 = m$ will immediately provide the following expression for $\underline{q}_i$:

$$\underline{q}_i = \begin{bmatrix} k_1^{(i)} \\ k_2^{(i)} \\ \cdot \\ \cdot \\ \cdot \\ k_{n_i-1}^{(i)} \end{bmatrix} \text{, where } n_i-1 = m \text{ .} \qquad (2.35)$$

It will prove useful to extend as follows a set of equations from (2.30) with $n_i-1 < m$:

$$\begin{bmatrix} \cdot \\ \cdot \\ \cdot \\ \underline{k}_m^{'}\underline{F}^{n_i-1} \\ \cdot \\ \cdot \\ \cdot \\ \underline{k}_m^{'}\underline{F}^{m-1} \end{bmatrix} \underline{q}_i =: \begin{bmatrix} \cdot \\ \cdot \\ \cdot \\ k_{n_i}^{*(i)} \\ \cdot \\ \cdot \\ \cdot \\ k_m^{*(i)} \end{bmatrix} .$$

Using Eq. (2.33), the following solution will then be obtained for $\underline{q}_i$:

$$\underline{q}_i = \begin{bmatrix} k_1^{(i)} \\ \cdot \\ \cdot \\ \cdot \\ k_{n_i-1}^{(i)} \\ k_{n_i}^{*(i)} \\ \cdot \\ \cdot \\ \cdot \\ k_m^{*(i)} \end{bmatrix} \quad , \text{ where } n_i-1 < m \; . \qquad (2.36)$$

The quantities

$$k_j^{*(i)} = \underline{k}_m'\underline{F}^{j-1}\underline{q}_i \quad , \quad j = n_i, \ldots m \qquad (2.37)$$

can be arbitrary selected as regards the requirement (2.9). They do not influence the nominal command behavior of the feedback system and will be defined only at a later point in the sensitivity design (section 2.5).

The abbreviation (2.29) coincides with (2.37) for $j = n_i$.

2.2.6 FREE PARAMETERS FOR THE CONTROLLER DESIGN

The requirement (2.9) will be satisfied where the r partial controllers comprising the controller are able to fulfill the design relationships indicated in sections 2.2.4 and 2.2.5. This involves the occurrence of some more free parameters which can be employed for the remaining requirements specified in section 2.2.2. These are, for each partial controller, the "starred" variables contained in the vectors $\underline{q}_i$ according to Eq. (2.36) and the eigenvalues of $\underline{F}$ according to Eq. (2.31).

According to Eq. (2.37) a structural index n_i with $n_i-1 < m$ will possess "starred" variables of precisely $m-n_i+1$.

The total quantity of these will be obtained by adding up above all i:

$$\eta = \sum_{i=1}^{p} \{m - n_i + 1\} \quad .$$

On account of $\sum_i n_i = n$, it thus follows:

> that the quantity η of the free parameters present along with the eigenvalues in a partial controller of the order $m \geq n_o - 1$ will be

$$\eta = p(m+1) - n \quad . \tag{2.38}$$

2.2.7 STRUCTURE IN THE $\mathfrak{z}$-DOMAIN

This section deals with a compilation and brief derivation of the feedback system relationships in the frequency domain. Sub-section 2.2.8 is of particular importance for the subject in question.

2.2.7.1 COMMAND BEHAVIOR

Following a $\mathfrak{z}$-transformation, the following relationship will be obtained from Eqs. (2.8), (2.9) and (2.10) for the command behavior control input:

$$\underline{U}(z) = \underline{\rho}\ \underline{R}(z) - \underline{K}(\underline{I}_n z - \underline{\Phi} + \underline{H}\ \underline{K})^{-1}\underline{H}\ \underline{\rho}\ \underline{R}(z) \ .$$

By using the notations

$$\underline{\Gamma}_R(z) := \underline{I}_r \Delta_R(z) - \underline{K}\ \mathrm{adj}(\underline{I}_n z - \underline{\Phi} + \underline{H}\ \underline{K})\underline{H} \tag{2.39}$$

and

$$\Delta_R(z) := \det(\underline{I}_n z - \underline{\Phi} + \underline{H}\ \underline{K}) \tag{2.40}$$

we shall obtain the following interrelationship between command input and control input:

$$\underline{U}(z) = \frac{\underline{\Gamma}_R(z)}{\Delta_R(z)}\ \underline{\rho}\ \underline{R}(z) \quad . \tag{2.41}$$

After multiplication of Eq.(2.41) by the plant transfer function

$$\underline{G}(z) := \frac{\underline{Z}(z)}{\Delta(z)} + \underline{D} \qquad (2.42)$$

$$\text{wherein } \underline{Z}(z) = \underline{C}\ \mathrm{adj}(\underline{I}_n z - \underline{\Phi})\underline{H} \qquad (2.43)$$

$$\Delta(z) = \det(\underline{I}_n z - \underline{\Phi}) \qquad (2.44)$$

the following expression will be obtained for the behavior of the output variable:

$$\underline{Y}(z) = \underline{G}(z) \frac{\underline{r}_R(z)}{\Delta_R(z)} \ \underline{\rho}\ \underline{R}(z) . \qquad (2.45)$$

2.2.7.2 DISTURBANCE BEHAVIOR AND SENSITIVITY

Fig.4 can be obtained in the $\mathcal{Z}$-range by means of a $\mathcal{Z}$-Transformation of the state space models (2.7) or with the aid of Fig.3.

The following is applicable for the control input in Fig.4:

$$\underline{U}(z) = \underline{\rho}\ \underline{R}(z) - \left\{\underline{\phi}(z) - \underline{Z}_C(z)\underline{D}\right\} \underline{U}(z) - \underline{Z}_C(z)\underline{Y}(z)$$

with

$$\underline{\phi}(z) := \begin{bmatrix} \underline{k}_m^{'(1)}\left\{\underline{I}\ z - \underline{F}^{(1)}\right\}^{-1}\underline{S}_u^{(1)} \\ \vdots \\ \underline{k}_m^{'(r)}\left\{\underline{I}\ z - \underline{F}^{(r)}\right\}^{-1}\underline{S}_u^{(r)} \end{bmatrix}$$

$$\underline{Z}_C(z) := \begin{bmatrix} \underline{k}_m^{'(1)}\left\{\underline{I}\ z - \underline{F}^{(1)}\right\}^{-1}\underline{S}_y^{(1)} + \underline{k}_y^{'(1)} \\ \vdots \\ \underline{k}_m^{'(r)}\left\{\underline{I}\ z - \underline{F}^{(r)}\right\}^{-1}\underline{S}_y^{(r)} + \underline{k}_y^{'(r)} \end{bmatrix} \qquad (2.46)$$

i.e., the following is applicable:

$$\underline{U}(z) = \left\{\underline{I}_r + \underline{\phi}(z) - \underline{Z}_C(z)\underline{D}\right\}^{-1} \left\{\underline{\rho}\ \underline{R}(z) - \underline{Z}_C(z)\underline{Y}(z)\right\} .$$

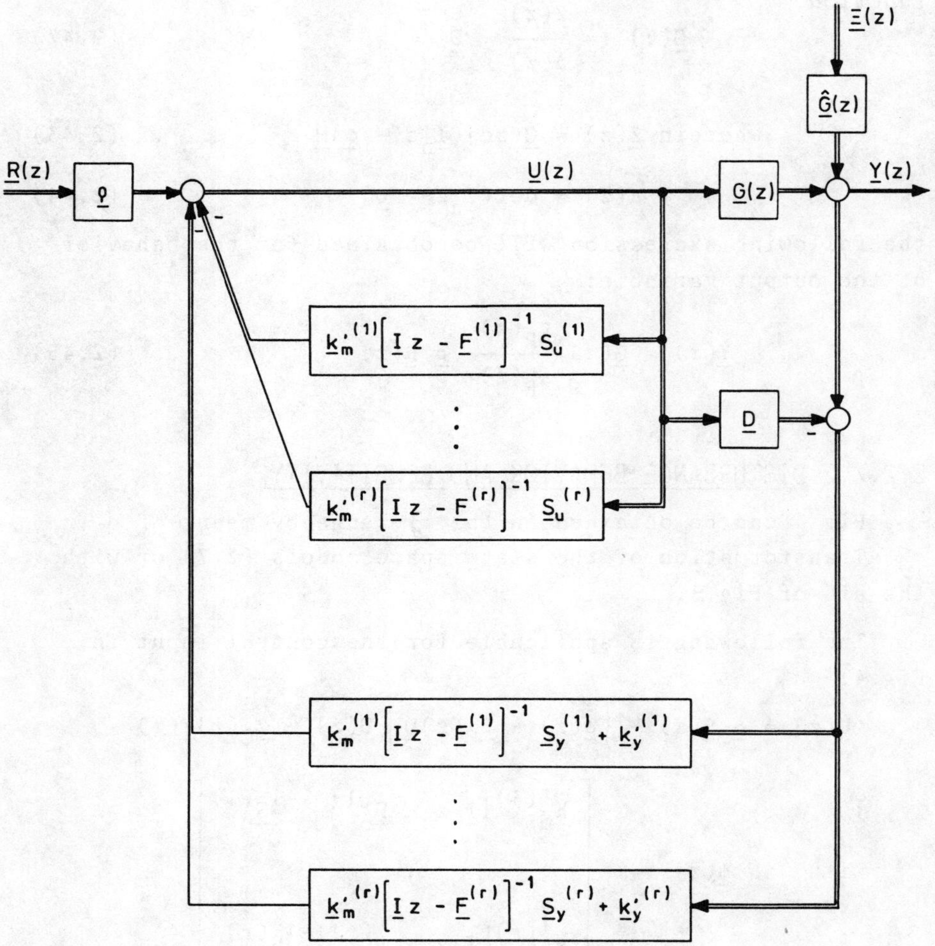

Fig.2.4: $\mathcal{Z}$-transformed feedback system of fig.2.3

By using a further abbreviation

$$\underline{\Gamma}_C(z) := \underline{I}_r + \underline{\phi}(z) - \underline{Z}_C(z)\underline{D} \qquad (2.47)$$

we finally obtain

$$\underline{U}(z) = \underline{\Gamma}_C^{-1}(z)\{\underline{\rho}\ \underline{R}(z) - \underline{Z}_C(z)\underline{Y}(z)\}$$

and, accordingly, Fig.5 as shown below (previously presented as Fig.2 in the Introduction) which, besides Fig.4, represents a further summarizing of the feedback system within the frequency domain.

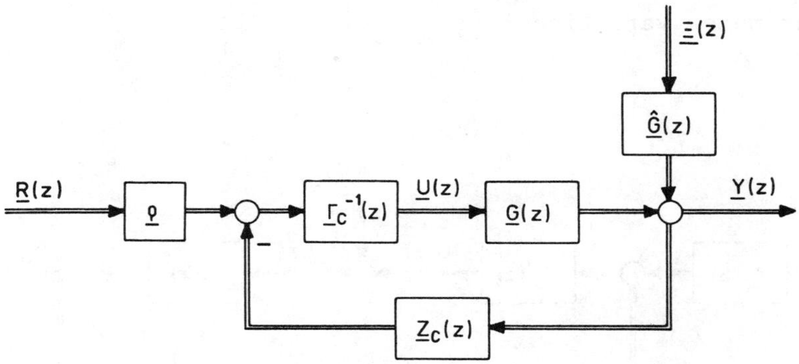

Fig.2.5: Structure of the disturbed feedback system in the frequency domain

2.2.7.3 GENERAL DISTURBANCE

The transfer function of the disturbance input $\underline{\Xi}(z)$ to the output of the plant is furnished by (cf.Fig.1)

$$\underline{\hat{G}}(z) = \underline{C}(\underline{I}_n z - \underline{\Phi})^{-1}\underline{E}_1 + \underline{E}_2 \quad . \tag{2.48}$$

With the aid of Fig.5 we then ascertain the following for the output behavior:

$$\underline{Y}(z) = \underline{S}(z)\underline{\hat{G}}(z)\underline{\Xi}(z) \tag{2.49}$$

with the inverse "effect of control"

$$\underline{S}(z) = \{\underline{I}_p + \underline{G}(z)\underline{\Gamma}_C^{-1}(z)\underline{Z}_C(z)\}^{-1} \quad .$$

In section 2.2.7.2.2 it will be explained that $\underline{S}(z)$ is of significance not only for the disturbance behavior but also for the sensitivity behavior with regard to parameter variations.

2.2.7.4 INPUT DISTURBANCES

If the disturbance input is applied at the plant input ($\underline{E}_1 = \underline{H}$, $\underline{E}_2 = \underline{O}$), then we shall obtain the following with $\hat{\underline{G}}(z) \equiv \underline{G}(z)$ from Eq.(2.49):

$$\underline{Y}(z) = \underline{S}(z)\underline{G}(z)\underline{\Xi}(z) \quad . \tag{2.50}$$

We should examine Fig.6 in order to deal with the nominal case (no parameter variations).

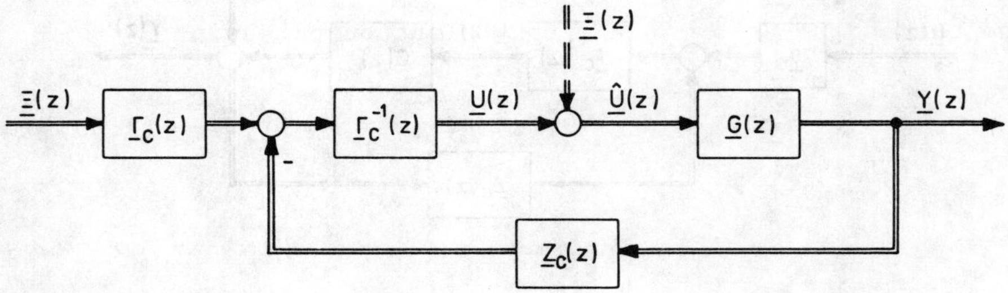

Fig. 2.6: Transformed disturbance in feedback system

The disturbance input $\underline{\Xi}(z)$ has been transformed at the point of application of the command input (cf. Fig.5). This enables Eq.(2.41) to be employed:

$$\hat{\underline{U}}(z) = \frac{\underline{\Gamma}_R(z)}{\Delta_R(z)} \underline{\Gamma}_C(z)\underline{\Xi}(z) \quad , \tag{2.51}$$

i.e.

$$\underline{U}(z) = \left\{ \frac{\underline{\Gamma}_R(z)}{\Delta_R(z)} \underline{\Gamma}_C(z) - \underline{I}_r \right\} \underline{\Xi}(z) \ . \tag{2.52}$$

The output behavior ensues from Eq.(2.51) by means of multiplication by the plant transfer function (2.42) (cf.Fig.6):

$$\underline{Y}(z) = \underline{G}(z) \frac{\underline{\Gamma}_R(z)}{\Delta_R(z)} \underline{\Gamma}_C(z)\underline{\Xi}(z) \ . \tag{2.53}$$

Thus, in the nominal case, we shall also obtain the relationship (2.53) along with Eq.(2.50).

2.2.7.5 OUTPUT DISTURBANCES

With $\underline{E}_1 = \underline{O}$, i.e. $\hat{\underline{G}}(z) \equiv \underline{E}_2$, the following will be obtained from Eq.(2.49) for the behavior with regard to output disturbances:

$$\underline{Y}(z) = \underline{S}(z)\underline{E}_2\underline{\Xi}(z) \quad . \tag{2.54}$$

If we interpret the output disturbance $\underline{\Xi}(z)$ as a negative command input $\underline{R}(z)$, then we shall obtain the following for the command behavior of the true plant output using $\underline{E}_2 = -\underline{I}_p$ (cf.Fig.7):

$$\underline{Y}(z) = (\underline{I}_p - \underline{S}(z))\underline{R}(z) \ . \tag{2.55}$$

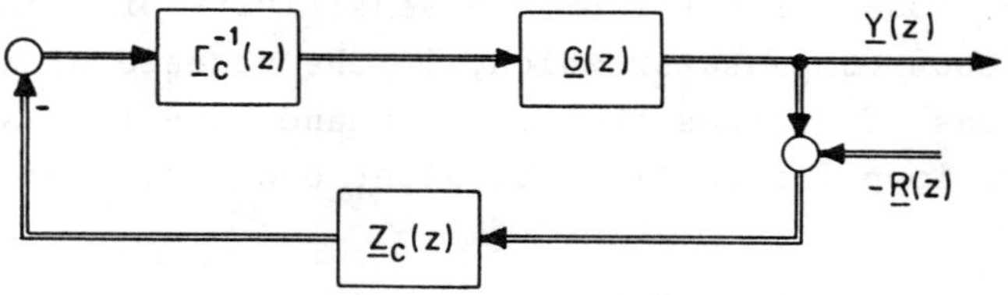

Fig. 2.7: Transformed command input in feedback system

We shall also obtain the following relationship for the nominal case by means of transformation of $\underline{R}(z)$ at the point of application of the command input in Fig.5 and by application of Eq.(2.45):

$$\underline{Y}(z) = \underline{G}(z) \frac{\underline{\Gamma}_R(z)}{\Delta_R(z)} \underline{Z}_C(z)\underline{R}(z) \quad . \tag{2.56}$$

It will become clear both from Eq.(2.55) and from Eq.(2.56) that it is not possible to define the nominal command behavior separately from the sensitivity design by employing the command input at the plant output. Therefore, we shall not give further attention to this method.

2.2.8 SENSITIVITY

The comparison sensitivity function [2] acts as a measurement of the sensitivity of the output variable for the command behavior with regard to parameter variations in the plant:

$$\underline{S}(z) = \{ \underline{I}_p + \underline{L}_p(z)\}^{-1} \qquad (2.57)$$

with (cf.Fig.5)

$$\underline{L}_p(z) = \underline{G}(z)\underline{\Gamma}_C^{-1}(z)\underline{Z}_C(z) \quad .$$

It indicates the factor by which the sensitivity of the closed control loop differs from the sensitivity of a corresponding open loop compensator which, in the absence of parameter variations, furnishes the same command behavior as the closed control loop (nominally equivalent open loop compensation):

$$\underline{E}_R(z) = \underline{S}(z)\underline{E}_S(z) \quad . \qquad (2.58)$$

Here, $\underline{E}_R(z)$ and $\underline{E}_S(z)$ are the output variable deviations determined by parameter variations for the closed control loop and for the nominally equivalent open loop compensator.

Should the changes in the plant transfer function be merely infinitesimal, then in Eq.(2.57) the actual transfer function would be transformed into the nominal transfer function of the plant. Eq.(2.57) could then be perceived as the generalization of Bode's sensitivity function to the multivariable case.

$\underline{S}(z)$ has already occurred in section 2.2.7.3 during the description of the disturbance behavior. By comparing the expressions (2.50) and (2.53) and (2.55) and (2.56) we shall also obtain the following two relationships for the sensitivity function in the nominal case:

$$\underline{S}(z)\underline{G}(z) = \underline{G}(z) \frac{\underline{r}_R(z)}{\Delta_R(z)} \underline{r}_C(z) \tag{2.59}$$

$$\underline{S}(z) = \underline{I}_p - \underline{G}(z) \frac{\underline{r}_R(z)}{\Delta_R(z)} \underline{Z}_C(z) \,. \tag{2.60}$$

2.2.9 SIGNIFICANCE OF THE PLANT ZEROS

If z_o is a (multiple) zero of the plant, lying on the unit circle or outside of it, then the following equation will follow from Eq.(2.60) for the nominal sensitivity function:

$$\underline{S}(z_o) = \underline{I}_p \quad , \tag{2.61}$$

since the zeros of $\Delta_R(z)$ and the pole positions of $\underline{Z}_C(z)$, i.e. the controller eigenvalues, lie in the unit circle and thus the plant zero in (2.60) cannot be cancelled out.

Consequently, it is not possible for the nominal sensitivity function at position z_o to be influenced by the controller design.

It also follows from Eq.(2.59) that

$$\underline{r}_C(z_o) \neq \underline{0} \tag{2.62}$$

must be applicable. This is clear from the fact that, taking into consideration Eq.(2.61), no zero can appear on the right-hand side of Eq.(2.59) which is of a higher order than on the left-hand side.

2.3 REQUIREMENTS REGARDING SENSITIVITY BEHAVIOR AND DISTURBANCE BEHAVIOR

With the aid of the feedback system relationships within the frequency domain as described in section 2.2.7, the remaining requirements made in section 2.2.2 can now be formulated in a more concrete manner and thereafter converted into requirements regarding the controller transfer functions.

If we formulate a vector norm from both sides of Eq.(2.58)

$$||\underline{E}_R(z)|| = ||\underline{S}(z)\underline{E}_S(z)|| \quad ,$$

and introduce for the sensitivity function $\underline{S}(z)$ a matrix norm suited to the vector norm, then

$$||\underline{E}_R(z)|| \leq ||\underline{S}(z)||\,||\underline{E}_S(z)|| \quad . \tag{2.63}$$

The control is less sensitive than an open loop compensation in respect of parameter variations when the following is applicable in the frequency range in question:

$$||\underline{S}(e^{j\varphi})|| < 1 \quad . \tag{2.64}$$

If we place a larger quantity q of frequency points $\varphi_1 \dots \varphi_q$ over the frequency range in question, then the following possible requirement will be obtained from (2.64):

$$\sum_{i=1}^{q} \alpha^2(\varphi_i)||\underline{S}(e^{j\varphi_i})||^2 \rightarrow \text{as small as possible.} \tag{2.65}$$

The aim to be achieved by the requirements (2.65) is a uniform minimization of $||S||$ in the frequency range $[\varphi_1, \varphi_q]$. The addition of the positive scalars $\alpha(\varphi_i)$ enables a differentiated weighting of individual sub-ranges.

Should it be desired not to obtain a minimization over a frequency range but, instead, to have insensitivity at individual frequencies, then the following requirement shall apply in place of (2.65):

$$\underline{S}(e^{j\varphi_i}) = \underline{0} \quad \text{for some } \varphi_i \quad . \tag{2.66}$$

Furthermore, it is possible to request the minimization over a frequency range in accordance with (2.65) using the secondary condition (2.66); eg. minimization of the sensitivity function within the band width of the command signals and, additionally, $\underline{S}(1) \overset{!}{=} \underline{0}$ for stationary precision.

With the aid of Eq.(2.49) the following requirement can be indicated for behavior in respect of disturbance inputs:

$$\sum_{i=1}^{q} \alpha^2(\varphi_i)||\underline{S}(e^{j\varphi_i})||^2||\hat{\underline{G}}(e^{j\varphi_i})||^2 \longrightarrow \text{as small as possible.} \quad (2.67)$$

If we imagine the $||\hat{\underline{G}}(e^{j\varphi_i})||$ to be contained within the $\alpha(\varphi_i)$, then (2.67) will likewise represent a requirement of type (2.65).

The requirements (2.65) and (2.66) must now be converted into requirements regarding the controller transfer functions $\underline{r}_C(z)$ and $\underline{Z}_C(z)$.

2.3.1 SENSITIVITY FUNCTION

2.3.1.1 CASE p = r

Should the plant posses the same number of inputs and outputs, then the plant transfer matrix $\underline{G}(z)$ will be quadratic. If we assume regularity of $\underline{G}(e^{j\varphi_i})$, then the requirement (2.66) will be identical with

$$\underline{r}_C(e^{j\varphi_i}) = \underline{0} \quad \text{for some } \varphi_i \text{ ,} \qquad (2.68)$$

which, for the nominal case, ensues directly from Eq.(2.59); otherwise, it will follow from the general expression (2.57) of the sensitivity function.

The relationship (2.59) will run as follows after multiplication by the inverse plant transfer matrix:

$$\underline{S}(z) = \underline{G}(z) \frac{\underline{r}_R(z)}{\Delta_R(z)} \underline{r}_C(z)\underline{G}^{-1}(z) .$$

Should $\underline{G}^*(z) = \underline{G}^{-1}(z)$, then the following will be applicable upon utilization of the Euclidean norm:

$$||\underline{S}(z)|| = \left\| \frac{\underline{r}_R(z)}{\Delta_R(z)} \underline{r}_C(z) \right\| \quad ; \qquad (2.69)$$

otherwise, the equality sign in (2.69) will be applicable only approximately. On account of the inequation $||\underline{A}\ \underline{B}|| \leq ||\underline{A}||\ ||\underline{B}||$ it should, however, be possible for the following to apply in most cases:

$$||\underline{S}(z)|| \leq \left\| \frac{\underline{r}_R(z)}{\Delta_R(z)} \right\| \; ||\underline{r}_C(z)|| \quad . \qquad (2.70)$$

(2.70) though, will always be applicable where the factor $||\underline{G}(z)||\ ||\underline{G}(z)^{-1}||$ is added on the right-hand side.

In view of (2.65),it is therefore practical to require the following:

$$\sum_{i=1}^{q} \alpha^2(\varphi_i) \left\| \frac{\underline{r}_R(e^{j\varphi_i})}{\Delta_R(e^{j\varphi_i})} \right\|^2 ||\underline{r}_C(e^{j\varphi_i})||^2 \longrightarrow \text{as small as possible.} \qquad (2.71)$$

2.3.1.2 CASE $p > r$

Since the right-hand matrix product in Eq.(2.60) is not now regular and cannot therefore accept the value of the unit matrix, it follows from Eq.(2.60) that the sensitivity function is unable to disappear and that the requirement (2.66) thus cannot be satisfied.

Consequently, the sensitivity design will be limited to r outputs and to the corresponding r rows of the sensitivity matrix, cf. Eq.(2.58). If we accept, without loss of generality, that we are dealing with the first r outputs and divide

$\underline{S}(z)$ as follows:

$$\underline{S}(z) = \underbrace{\begin{bmatrix} \underline{S}_1(z) \\ \\ \underline{S}_2(z) \end{bmatrix}}_{p} \begin{matrix} \}\ r \\ \\ \}\ p-r \end{matrix} \quad , \tag{2.72}$$

then, in place of (2.66), the following has to be required:

$$\underline{S}_1(e^{j\varphi_i}) = \underline{0} \qquad \text{for some} \quad \varphi_i \ . \tag{2.73}$$

If we also divide the plant transfer matrix into

$$\underline{G}(z) = \underbrace{\begin{bmatrix} \underline{G}_1(z) \\ \\ \underline{G}_2(z) \end{bmatrix}}_{r} \begin{matrix} \}\ r \\ \\ \}\ p-r \end{matrix} \quad , \tag{2.74}$$

then the first r rows of Eq.(2.59) will read as follows:

$$\underline{S}_1(z)\underline{G}(z) = \underline{G}_1(z) \frac{\underline{r}_R(z)}{\Delta_R(z)} \underline{r}_C(z) \quad . \tag{2.75}$$

It ensues from this that (where $\underline{G}_1$ and $\underline{r}_R$ are regular) (2.68) will indeed follow from (2.73). The inverse of this, however, is not applicable because $\underline{G}(z)$ does not have independent rows. Consequently, the validity of (2.68) is insufficient in the case $p > r$.

If we now break down $\underline{S}_1(z)$ further into

$$\underline{S}_1(z) = [\underbrace{\underline{S}_{11}(z)}_{r} \quad \underbrace{\underline{S}_{12}(z)}_{p-r}] \ \}\ r \tag{2.76}$$

and also make a corresponding division of $\underline{Z}_C(z)$:

$$\underline{Z}_C(z) = [\underbrace{\underline{Z}_{C1}(z)}_{r} \quad \underbrace{\underline{Z}_{C2}(z)}_{p-r}] \ \}\ r \quad , \tag{2.77}$$

then the first r rows of Eq.(2.60) will be transformed into

$$[\underline{S}_{11}(z)\ \underline{S}_{12}(z)] = [\underline{I}_r \quad \underline{O}] - \underline{G}_1(z)\frac{\underline{\Gamma}_R(z)}{\Delta_R(z)}[\underline{Z}_{C1}(z) \quad \underline{Z}_{C2}(z)] ,$$

wherefrom the following will be obtained for $\underline{S}_{12}(z)$:

$$\underline{S}_{12}(z) = -\underline{G}_1(z)\frac{\underline{\Gamma}_R(z)}{\Delta_R(z)}\underline{Z}_{C2}(z) \quad . \qquad (2.78)$$

It ensues from Eq.(2.78) that with

$$\underline{Z}_{C2}(e^{j\varphi_i}) = \underline{O} \qquad (2.79)$$

the following would also be applicable:

$$\underline{S}_{12}(e^{j\varphi_i}) = \underline{O} \quad . \qquad (2.80)$$

Eq.(2.75) which with (2.74) and (2.76) has the following appearance:

$$[\underline{S}_{11}(z) \quad \underline{S}_{12}(z)]\begin{bmatrix}\underline{G}_1(z)\\ \underline{G}_2(z)\end{bmatrix} = \underline{G}_1(z)\frac{\underline{\Gamma}_R(z)}{\Delta_R(z)}\underline{\Gamma}_C(z) ,$$

must then provide the following relationship on taking Eq. (2.80) into consideration:

$$\underline{S}_{11}(z)\underline{G}_1(z) = \underline{G}_1\frac{\underline{\Gamma}_R(z)}{\Delta_R(z)}\underline{\Gamma}_C(z) \quad . \qquad (2.81)$$

Since $\underline{G}_1$ is quadratic, $\underline{S}_{11}$ also would disappear with $\underline{\Gamma}_C$ (where $\underline{G}_1$ is regular).

It is thus deduced for nominal cases that the requirements (2.68) and (2.79) are equivalent to the requirement (2.73). In order to demonstrate this result for general cases also, we shall introduce Eq.(2.79) into Eq.(2.57):

$$\underline{S}^{-1} = \underline{I}_p + \begin{bmatrix}\underline{G}_1\\ \underline{G}_2\end{bmatrix}\underline{\Gamma}_C^{-1}[\underline{Z}_{C1} \quad \underline{O}] \quad .$$

This will provide

$$\underline{S}^{-1} = \begin{bmatrix} \underline{I}_r + \underline{G}_1\underline{r}_C^{-1}\underline{Z}_{C1} & \underline{0} \\ \underline{G}_2\underline{r}_C^{-1}\underline{Z}_{C1} & \underline{I}_{p-r} \end{bmatrix}$$

and the formation of the inverse matrix will provide the following expression for the sensitivity function:

$$\underline{S} = \begin{bmatrix} (\underline{I}_r + \underline{G}_1\underline{r}_C^{-1}\underline{Z}_{C1})^{-1} & \underline{0} \\ -\underline{G}_2\underline{r}_C^{-1}\underline{Z}_{C1}(\underline{I}_r + \underline{G}_1\underline{r}_C^{-1}\underline{Z}_{C1})^{-1} & \underline{I}_{p-r} \end{bmatrix} \quad (2.82)$$

Since $\underline{G}_1$ is quadratic, it follows (where $\underline{G}_1$ is regular) from Eq.(2.82) that the first r rows of the sensitivity function will also disappear with $\underline{r}_C = \underline{0}$.

It will thus have been demonstrated that in the case $p > r$ the requirement (2.79)

$$\underline{Z}_{C2}(e^{j\varphi_i}) = \underline{0} \qquad \text{for some} \quad \varphi_i \qquad (2.79)$$

has added itself to the requirement (2.68), with the result that (2.73) will now be applicable.

The following expression will be obtained for $\underline{S}_1(z)$ with Eq.(2.81) and Eq.(2.78) (where $\underline{G}_1$ is regular):

$$\underline{S}_1(z) = \left[\underline{G}_1(z)\,\frac{\underline{r}_R(z)}{\Delta_R(z)}\,\underline{r}_C(z)\underline{G}_1^{-1}(z) \;\middle|\; -\underline{G}_1(z)\,\frac{\underline{r}_R(z)}{\Delta_R(z)}\,\underline{Z}_{C2}(z) \right] .$$

The requirement (2.71) is therefore completed by means of the requirement

$$\sum_{i=1}^{q} \alpha^2(\varphi_i) \left\| \underline{G}_1(e^{j\varphi_i})\,\frac{\underline{r}_R(e^{j\varphi_i})}{\Delta_R(e^{j\varphi_i})} \right\|^2 ||\underline{Z}_{C2}(e^{j\varphi_i})||^2$$

$$\longrightarrow \text{as small as possible.} \quad (2.83)$$

2.3.1.3 CASE $p < r$

Should the plant possess a smaller number of outputs than inputs, then the requirement (2.66) will be able to be satisfied merely by the disappearance of p columns from $\underline{\Gamma}_C(e^{j\varphi_i})$.

In order to demonstrate this first of all for the nominal case, we shall subdivide $\underline{G}(z)$ and $\underline{\Gamma}_C(z)$ as follows:

$$\underline{G}(z) = [\underbrace{\underline{G}_1(z)}_{p} \quad \underbrace{\underline{G}_2(z)}_{r-p}] \} p$$

$$\underline{\Gamma}_C(z) = [\underbrace{\underline{\Gamma}_{C1}(z)}_{p} \quad \underbrace{\underline{\Gamma}_{C2}(z)}_{r-p}] \} r \quad .$$

The first p columns of Eq.(2.59) will thus read as follows:

$$\underline{S}(z)\underline{G}_1(z) = \underline{G}(z) \frac{\underline{\Gamma}_R(z)}{\Delta_R(z)} \underline{\Gamma}_{C1}(z) \quad . \tag{2.84}$$

It ensues from Eq.(2.84) that where $\underline{G}_1(e^{j\varphi_i})$ is regular, $\underline{S}(e^{j\varphi_i})$ will also disappear with $\underline{\Gamma}_{C1}(e^{j\varphi_i})$.

The same result is also obtained for the general sensitivity function (2.57), this being because the (p x p) matrix $\underline{S}(e^{j\varphi_i})$ in accordance with (2.57) will disappear (where $\underline{G}\,\underline{\Gamma}_C^{-1}\underline{Z}_C$ is regular) as soon as p columns alone disappear from $\underline{\Gamma}_C(e^{j\varphi_i})$.

2.3.2 DISTURBANCE TRANSFER FUNCTION

As explained in section 2.3.1.2, in the case $p > r$ it is possible only for r rows of the sensitivity function to disappear and, accordingly, only the corresponding r rows of the disturbance transfer function $\underline{S}(z)\hat{\underline{G}}(z)$. Should, however, an input disturbance, $\hat{\underline{G}}(z) \equiv \underline{G}(z)$, be involved, then in this case the whole disturbance transfer matrix will disappear, this being because, taking into consideration Eq.(2.74), the following will be obtained with Eq.(2.82):

$$\underline{S}\,\underline{G} = \begin{bmatrix} (\underline{I}_r + \underline{G}_1\underline{r}_C^{-1}\underline{Z}_{C1})^{-1}\underline{G}_1 \\ \\ -\underline{G}_2\underline{r}_C^{-1}\underline{Z}_{C1}(\underline{I}_r + \underline{G}_1\underline{r}_C^{-1}\underline{Z}_{C1})^{-1}\underline{G}_1 + \underline{G}_2 \end{bmatrix} ,$$

and, with the assistance of the general matrix relationship $\underline{A}(\underline{I} + \underline{B}\,\underline{A})^{-1}\underline{B} = \underline{I} - (\underline{I} + \underline{A}\,\underline{B})^{-1}$, the following will ensue therefrom:

$$\underline{S}\,\underline{G} = \begin{bmatrix} (\underline{I}_r + \underline{G}_1\underline{r}_C^{-1}\underline{Z}_{C1})^{-1}\underline{G}_1 \\ \\ -\underline{G}_2\{\underline{I}_r - (\underline{I}_r + \underline{r}_C^{-1}\underline{Z}_{C1}\underline{G}_1)^{-1}\} + \underline{G}_2 \end{bmatrix}$$

and, accordingly,

$$\underline{S}\,\underline{G} = \begin{bmatrix} (\underline{I}_r + \underline{G}_1\underline{r}_C^{-1}\underline{Z}_{C1})^{-1}\underline{G}_1 \\ \\ \underline{G}_2(\underline{I}_r + \underline{r}_C^{-1}\underline{Z}_{C1}\underline{G}_1)^{-1} \end{bmatrix} . \tag{2.85}$$

With $\underline{r}_C = \underline{0}$ the occurring inverse matrices $(\underline{I}_r + \ldots)^{-1}$ will disappear (where $\underline{G}_1$ is regular) and, consequently, the whole disturbance transfer matrix $\underline{S}\,\underline{G}$.

A further particiuliarity regarding the input disturbance behavior is the fact that in the nominal case the requirement (2.79) (occurring in the case $p > r$) can be dispensed with. This becomes immediately clear from Eq.(2.53).

2.4 SPECIAL REPRESENTATIONS OF THE CONTROLLER TRANSFER FUNCTIONS

In order to be able to satisfy the requirements made in the previous section regarding the controller transfer functions $\underline{r}_C(z)$ and $\underline{Z}_C(z)$, special representations shall first of all be derived for $\underline{Z}_C(z)$ (section 2.4.1) and $\underline{r}_C(z)$ (section 2.4.2).

For this purpose the following two auxiliary relationships will prove useful. With the exception of the limitations $m \geq n-1$ and

$$k_j = -\begin{bmatrix} k_{j-m}, & \dots, & k_{j-1} \end{bmatrix} \begin{bmatrix} ß_o \\ ß_1 \\ \vdots \\ ß_{m-1} \end{bmatrix} \qquad \text{für } j > m \quad ,$$

they are applicable for arbitrary values of $ß_i$, k_i and h_i, these three notations having been selected with regard to the future application of the auxiliary relationships.

Auxiliary relationship 1:

$$\begin{bmatrix} 1 & z & \dots & z^{m-1} \end{bmatrix} \begin{bmatrix} ß_1 & ß_2 & \dots & ß_{m-1} & 1 \\ ß_2 & ß_3 & & \cdot & \\ \vdots & & \cdot & & \\ ß_{m-1} & & & \underline{O} & \\ 1 & & & & \end{bmatrix} \begin{bmatrix} k_1 & \dots & k_{n-1} & k_n \\ k_2 & \dots & k_n & k_{n+1} \\ \vdots & & & \vdots \\ k_m & & & k_{n+m-1} \end{bmatrix} \begin{bmatrix} h_1 \\ h_2 \\ \vdots \\ h_n \end{bmatrix}$$

$$= l_1(z)k_1 + l_2(z)k_2 + \dots + l_m(z)k_m$$

with (2.86)

$$l_s(z) = \sum_{\mu=1}^{\bar{s}} \sum_{\nu=s}^{m} h_\mu ß_\nu z^{\mu+\nu-s-1} - \sum_{\mu=o}^{s-1} \sum_{\nu=s+1}^{n} ß_\mu h_\nu z^{\mu+\nu-s-1}$$

$$\bar{s} = \min(s,n);\ s = 1,\dots m$$

Auxiliary relationship 2:

$$\begin{bmatrix} 1 & z & \dots & z^{m-1} \end{bmatrix} \begin{bmatrix} \beta_1 & \beta_2 & \dots & \beta_{m-1} & 1 \\ \beta_2 & \beta_3 & & \cdot^{\cdot^{\cdot}} & \\ \vdots & & \cdot^{\cdot^{\cdot}} & & \\ \beta_{m-1} & & & \underline{0} & \\ 1 & & & & \end{bmatrix} \begin{bmatrix} k_{n+1} \\ k_{n+2} \\ \vdots \\ k_{n+m} \end{bmatrix} =$$

$$= \hat{l}_1(z)k_1 + \hat{l}_2(z)k_2 + \dots + \hat{l}_m(z)k_m \qquad (2.87)$$

with

$$\hat{l}_s(z) = \left. \begin{cases} -\sum\limits_{\nu=o}^{s-1} \beta_\nu z^{\nu+n-s} & \text{for } s = 1,\dots n \\ +\sum\limits_{\nu=s}^{m} \beta_\nu z^{\nu+n-s} & \text{for } s = n+1,\dots m \end{cases} \right\} m \geq n$$

$$\hat{l}_s(z) = \beta_{s-1} \sum_{\nu=o}^{m-1} \beta_\nu z^\nu - \sum_{\nu=o}^{s-2} \beta_\nu z^{\nu+n-s} \qquad m = n-1$$

$$s = 1,\dots m$$

These auxiliary relationships can be located by multiplying out the matrices and ordering the individual terms accordingly.

2.4.1 PREPARATION OF $\underline{Z}_C(z)$ FOR THE SENSITIVITY DESIGN

The controller transfer function $\underline{Z}_C(z)$ according to Eq. (2.46) is transformed with

$$(\underline{I}\, z - \underline{F}^{(i)})^{-1} = \frac{\text{adj}(\underline{I}\, z - \underline{F}^{(i)})}{\det(\underline{I}\, z - \underline{F}^{(i)})} \qquad (2.88)$$

into

$$\underline{Z}_C(z) = \begin{bmatrix} \underline{k}_y^{'(1)} + \underline{k}_m^{'(1)} \dfrac{\mathrm{adj}(\underline{I}\ z - \underline{F}^{(1)})}{\det(\underline{I}\ z - \underline{F}^{(1)})} \underline{S}_y^{(1)} \\ \cdot \\ \cdot \\ \cdot \\ \underline{k}_y^{'(r)} + \underline{k}_m^{'(r)} \dfrac{\mathrm{adj}(\underline{I}\ z - \underline{F}^{(r)})}{\det(\underline{I}\ z - \underline{F}^{(r)})} \underline{S}_y^{(r)} \end{bmatrix} .$$

Thus, for $\underline{Z}_C(z)$ we have the experession

$$\underline{Z}_C(z) = \begin{bmatrix} \dfrac{\underline{z}_1^{'}(z)}{\Delta_B^{(1)}(z)} \\ \cdot \\ \cdot \\ \cdot \\ \dfrac{\underline{z}_r^{'}(z)}{\Delta_B^{(r)}(z)} \end{bmatrix} \qquad (2.89)$$

with the polynomial vectors

$$\underline{z}_i^{'}(z) = \underline{k}_y^{'(i)} \Delta_B^{(i)}(z) + \underline{k}_m^{'(i)} \mathrm{adj}(\underline{I}\ z - \underline{F}^{(i)}) \underline{S}_y^{(i)} \qquad (2.90)$$

and the characteristic polynomials

$$\Delta_B^{(i)}(z) = \det(\underline{I}\ z - \underline{F}^{(i)}) \qquad (2.91)$$

wherein $i = 1, \ldots r$.

For the sake of improved clarity, the index (i) shall henceforth be omitted (as previously in sections 2.2.4 to 2.2.6) for the controller matrices and the characteristic polynomials $\Delta_B^{(i)}(z)$. This is possible because all partial controllers are designed in the same manner.

Using the summation representation

$$\mathrm{adj}(\underline{I}\ z - \underline{F}) = \sum_{\nu=o}^{m-1} z^{\nu} \sum_{\mu=1}^{m-\nu} \beta_{\mu+\nu} \underline{F}^{\mu-1}$$

for the adjunct matrix of $\underline{I}z$-$\underline{F}$ (see Eq.(2.31) on account of β_{μ})

we shall obtain the following from Eq.(2.90):

$$\underline{z}_i'(z) = \underline{k}_y'\Delta_B(z) + \sum_{\nu=o}^{m-1} z^{\nu} \sum_{\mu=1}^{m-\nu} \beta_{\mu+\nu}\underline{k}_m'\underline{F}^{\mu-1}\underline{S}_y$$

or, in matrix form,

$$\underline{z}_i'(z) = \underline{k}_y'\Delta_B(z) + \begin{bmatrix} 1 & z & \dots & z^{m-1} \end{bmatrix} \begin{bmatrix} \beta_1 & \beta_2 & \dots & \beta_{m-1} & 1 \\ \beta_2 & \beta_3 & & & \\ \vdots & & & \cdot^{\cdot^{\cdot}} & \\ \beta_{m-1} & & & \underline{0} & \\ 1 & & & & \end{bmatrix} \underline{\Lambda} \quad (2.92)$$

$$\text{with} \quad \underline{\Lambda} = \begin{bmatrix} \underline{k}_m' \\ \underline{k}_m'\underline{F} \\ \vdots \\ \underline{k}_m'\underline{F}^{m-1} \end{bmatrix} \underline{S}_y .$$

Taking into consideration the Eq.(2.26) for $\underline{S}_y$

$$\underline{S}_y = \tilde{\underline{Q}} \begin{bmatrix} \underline{a}_1 & \dots & \underline{a}_p \end{bmatrix} \underline{P}^{-1} - \begin{bmatrix} \underline{F}^{n_1}\underline{q}_1 & \dots & \underline{F}^{n_p}\underline{q}_p \end{bmatrix} \underline{P}^{-1} \quad (2.93)$$

the following is obtained for $\underline{\Lambda}$:

$$\underline{\Lambda} = \underline{\Lambda}_1 \begin{bmatrix} \underline{a}_1 & \dots & \underline{a}_p \end{bmatrix} \underline{P}^{-1} - \underline{\Lambda}_2 \underline{P}^{-1} \quad (2.94)$$

$$\text{with} \quad \underline{\Lambda}_1 = \begin{bmatrix} \underline{k}_m'\tilde{\underline{Q}} \\ \underline{k}_m'\underline{F}\tilde{\underline{Q}} \\ \vdots \\ \underline{k}_m'\underline{F}^{m-1}\tilde{\underline{Q}} \end{bmatrix}$$

$$\underline{\Lambda}_2 = \begin{bmatrix} \underline{k}_m'\underline{F}^{n_1}\underline{q}_1 & \dots & \underline{k}_m'\underline{F}^{n_p}\underline{q}_p \\ \underline{k}_m'\underline{F}^{n_1+1}\underline{q}_1 & \dots & \underline{k}_m'\underline{F}^{n_p+1}\underline{q}_p \\ \vdots & & \\ \underline{k}_m'\underline{F}^{n_1+m-1}\underline{q}_1 & \dots & \underline{k}_m'\underline{F}^{n_p+m-1}\underline{q}_p \end{bmatrix} .$$

Since $\tilde{\underline{Q}}$ has the form (2.25), the following is applicable:

$$\underline{\Lambda}_1 = \left[\begin{array}{cccc|c|ccc} \underline{k}_m'\underline{q}_1 & \underline{k}_m'\underline{F}\underline{q}_1 & \cdots & \underline{k}_m'\underline{F}^{n_1-1}\underline{q}_1 & & \underline{k}_m'\underline{q}_p & \cdots & \underline{k}_m'\underline{F}^{n_p-1}\underline{q}_p \\ \underline{k}_m'\underline{F}\underline{q}_1 & \underline{k}_m'\underline{F}^2\underline{q}_1 & \cdots & \underline{k}_m'\underline{F}^{n_1}\underline{q}_1 & \cdots & \underline{k}_m'\underline{F}\underline{q}_p & \cdots & \underline{k}_m'\underline{F}^{n_p}\underline{q}_p \\ & \vdots & & & & & \vdots & \\ \underline{k}_m'\underline{F}^{m-1}\underline{q}_1 & & \cdots & \underline{k}_m'\underline{F}^{m+n_1-2}\underline{q}_1 & & & \cdots & \underline{k}_m'\underline{F}^{m+n_p-2}\underline{q}_p \end{array}\right].$$

Taking into consideration Eq.(2.28) the following expressions will be given for $\underline{\Lambda}_1$ and $\underline{\Lambda}_2$:

$$\underline{\Lambda}_1 = \left[\begin{array}{ccc|c|ccc} k_1^{(1)} & \cdots \; k_{n_1-1}^{(1)} & k_{n_1}^{*(1)} & & k_1^{(p)} & \cdots \; k_{n_p-1}^{(p)} & k_{n_p}^{*(p)} \\ k_2^{(1)} & \cdots \; k_{n_1}^{*(1)} & k_{n_1+1}^{*(1)} & \cdots & k_2^{(p)} & \cdots \; k_{n_p}^{*(p)} & k_{n_p+1}^{*(p)} \\ & \ddots & \vdots & & & \ddots & \vdots \\ & & k_{n_1+m-1}^{*(1)} & & & & k_{n_p+m-1}^{*(p)} \end{array}\right] \quad (2.95)$$

$$\underline{\Lambda}_2 = \left[\begin{array}{c|c|c} k_{n_1+1}^{*(1)} & & k_{n_p+1}^{*(p)} \\ k_{n_1+2}^{*(1)} & \cdots & k_{n_p+2}^{*(p)} \\ \vdots & & \vdots \\ k_{n_1+m}^{*(1)} & & k_{n_p+m}^{*(p)} \end{array}\right], \quad (2.96)$$

wherein the abbreviations

$$k_j^{*(\zeta)} := \underline{k}_m'\underline{F}^{j-1}\underline{q}_\zeta \;, \; j = n_\zeta, n_\zeta+1, \ldots n_\zeta+m \; (\zeta = 1, \ldots p) \quad (2.97)$$

will have been used. The parameters (2.97) coincide with the parameters (2.37) as far as the index j = m. The "starred" variables with an index $j > m$ are, on the basis of the Cayley-Hamilton theorem, whereby the matrix $\underline{F}$ satisfies its own characteristic equation, determined by means of the variables which precede m in each case:

$$k_j^{*(\zeta)} = - \left[k_{j-m}^{(*)(\zeta)}, \ldots, k_{j-1}^{(*)(\zeta)} \right] \begin{bmatrix} \beta_o \\ \beta_1 \\ \vdots \\ \beta_{m-1} \end{bmatrix} , \quad j > m . \tag{2.98}$$

If we introduce the expressions (2.95) and (2.96) into Eq.(2.94) and subdivide the matrices $[\underline{a}_1 \ldots \underline{a}_p]\underline{P}^{-1}$ and $\underline{P}^{-1}$ in Eq.(2.94) as follows:

$$[\underline{a}_1 \ldots \underline{a}_p]\underline{P}^{-1} =: \begin{bmatrix} \underline{a}_1^{'(1)} \\ \vdots \\ \underline{a}_{n_1}^{'(1)} \\ --- \\ \vdots \\ --- \\ \underline{a}_1^{'(p)} \\ \vdots \\ \underline{a}_{n_p}^{'(p)} \end{bmatrix} \tag{2.99}$$

$$\underline{P}^{-1} =: \begin{bmatrix} \underline{p}_1' \\ \underline{p}_2' \\ \vdots \\ \underline{p}_p' \end{bmatrix} , \tag{2.100}$$

then $\underline{\Lambda}$ will be transformed into

$$\underline{\Lambda} = \begin{bmatrix} k_1^{(1)} & \cdots & k_{n_1-1}^{(1)} & k_{n_1}^{*(1)} \\ k_2^{(1)} & \cdots & k_{n_1}^{*(1)} & k_{n_1+1}^{*(1)} \\ & \cdot & & \cdot \\ & \cdot & & \cdot \\ \cdot & & & \cdot \\ & & & k_{n_1+m-1}^{*(1)} \end{bmatrix} \begin{bmatrix} \underline{a}_1^{'(1)} \\ \underline{a}_2^{'(1)} \\ \cdot \\ \cdot \\ \cdot \\ \underline{a}_{n_1}^{'(1)} \end{bmatrix} +$$

$$+ \cdots + \begin{bmatrix} k_1^{(p)} & \cdots & k_{n_p-1}^{(p)} & k_{n_p}^{*(p)} \\ k_2^{(p)} & \cdots & k_{n_p}^{*(p)} & k_{n_p+1}^{*(p)} \\ & \cdot & & \cdot \\ & \cdot & & \cdot \\ \cdot & & & \cdot \\ & & & k_{n_p+m-1}^{*(p)} \end{bmatrix} \begin{bmatrix} \underline{a}_1^{'(p)} \\ \underline{a}_2^{'(p)} \\ \cdot \\ \cdot \\ \cdot \\ \underline{a}_{n_p}^{'(p)} \end{bmatrix} -$$

$$- \begin{bmatrix} k_{n_1+1}^{*(1)} \\ k_{n_1+2}^{*(1)} \\ \cdot \\ \cdot \\ \cdot \\ k_{n_1+m}^{*(1)} \end{bmatrix} \underline{p}_1' - \cdots - \begin{bmatrix} k_{n_p+1}^{*(p)} \\ k_{n_p+2}^{*(p)} \\ \cdot \\ \cdot \\ \cdot \\ k_{n_p+m}^{*(p)} \end{bmatrix} \underline{p}_p' \quad .$$

By using the two auxiliary relationships (2.86) and (2.87) we are able to identify the following representation from the above for the right-hand addend in Eq.(2.92):

$$\begin{bmatrix} 1 & z & \dots & z^{m-1} \end{bmatrix} \begin{bmatrix} \beta_1 & \beta_2 & \dots & \beta_{m-1} & 1 \\ \beta_2 & \beta_3 & & & \\ \vdots & & & \cdot^{\cdot^{\cdot}} & \\ \beta_{m-1} & & & \underline{O} & \\ 1 & & & & \end{bmatrix} \underline{\Lambda} = \tag{2.101}$$

$$= \underline{l}_1^{'(1)}(z)k_1^{(1)} + \dots + \underline{l}_{n_1-1}^{'(1)}(z)k_{n_1-1}^{(1)} + \underline{l}_{n_1}^{'(1)}(z)k_{n_1}^{*(1)} + \dots + \underline{l}_m^{'(1)}(z)k_m^{*(1)}$$

$$\vdots$$

$$+ \underline{l}_1^{'(p)}(z)k_1^{(p)} + \dots + \underline{l}_{n_p-1}^{'(p)}(z)k_{n_p-1}^{(p)} + \underline{l}_{n_p}^{'(p)}(z)k_{n_p}^{*(p)} + \dots + \underline{l}_m^{'(p)}(z)k_m^{*(p)}$$

with

$$\underline{l}_s^{(\varkappa)}(z) = \sum_{\mu=1}^{\bar{s}} \sum_{\nu=s}^{m} \underline{a}_\mu^{(\varkappa)} \beta_\nu z^{\mu+\nu-s-1} - \sum_{\mu=o}^{s-1} \sum_{\nu=s+1}^{n_\varkappa} \beta_\mu \underline{a}_\nu^{(\varkappa)} z^{\mu+\nu-s-1} +$$

$$\left.\begin{aligned} &+ \sum_{\nu=o}^{s-1} \beta_\nu z^{\nu+n_\varkappa-s} \underline{p}_\varkappa \quad \text{for} \quad s = 1,\dots,n_\varkappa \\ &- \sum_{\nu=s}^{m} \beta_\nu z^{\nu+n_\varkappa-s} \underline{p}_\varkappa \quad \text{for} \quad s = n_\varkappa+1,\dots m \end{aligned}\right\} \quad m \geq n_\varkappa$$

$$- \beta_{s-1} \sum_{\mu=o}^{m-1} \beta_\mu z^\mu \underline{p}_\varkappa + \sum_{\mu=o}^{s-2} \beta_\mu z^{\mu+n_\varkappa-s} \underline{p}_\varkappa \qquad m = n_\varkappa - 1$$

$$\bar{s} = \min(s,n_\varkappa) \qquad s = 1,\dots m \qquad \varkappa = 1,\dots p \qquad m \geq n_o - 1$$

Consideration must now be given to the left-hand addend in Eq.(2.92). This will read as follows with Eq.(2.27):

$$\underline{k}_y' \Delta_B(z) = \left[k_{n_1}^{(1)}, k_{n_2}^{(2)}, \ldots, k_{n_p}^{(p)}\right] \underline{P}^{-1} \Delta_B(z)$$

$$- \left[k_{n_1}^{*(1)}, k_{n_2}^{*(2)}, \ldots, k_{n_p}^{*(p)}\right] \underline{P}^{-1} \Delta_B(z) \quad ,$$

and, with (2.100):

$$\underline{k}_y' \Delta_B(z) = \left[k_{n_1}^{(1)}, k_{n_2}^{(2)}, \ldots, k_{n_p}^{(p)}\right] \underline{P}^{-1} \Delta_B(z)$$
$$- k_{n_1}^{*(1)} \underline{p}_1' \Delta_B(z) - \ldots - k_{n_p}^{*(p)} \underline{p}_p' \Delta_B(z) \quad . \tag{2.102}$$

The purpose of the observation made below is to incorporate into Eq.(2.101) the second row of Eq.(2.102) i.e.
$- k_{n_\varkappa}^{*(\varkappa)} \underline{p}_\varkappa' \Delta_B(z)$, $\varkappa = 1,\ldots p$.
Accordingly, it is necessary to differentiate between the cases $m \geq n_\varkappa$ and $m = n_\varkappa - 1$.

In the case $m \geq n_\varkappa$, $k_{n_\varkappa}^{*(\varkappa)}$ can be freely selected (cf.section 2.2.5) and the coefficient $\underline{l}_{n_\varkappa}^{'(\varkappa)}$ in Eq.(2.101) is simply completed by the addend

$$- \underline{p}_\varkappa' \Delta_B(z) \quad .$$

It can be added to the expression

$$\sum_{\nu=0}^{s-1} \beta_\nu z^{\nu+n_\varkappa-s} \underline{p}_\varkappa \qquad \text{for} \quad s = n_\varkappa$$

$$= \sum_{\nu=0}^{n_\varkappa-1} \beta_\nu z^\nu \underline{p}_\varkappa$$

which will then be transformed into

$$- \underline{p}_\varkappa \, \Delta_B(z) + \sum_{\nu=0}^{n_\varkappa - 1} \beta_\nu z^\nu \underline{p}_\varkappa$$

$$= - \sum_{\nu=n_\varkappa}^{m} \beta_\nu z^\nu \underline{p}_\varkappa$$

$$= - \sum_{\nu=s}^{m} \beta_\nu z^{\nu+n_\varkappa-s} \underline{p}_\varkappa \qquad \text{with } s = n_\varkappa \quad .$$

The last three rows of $\underline{1}_s^{(\varkappa)}(z)$ in Eq.(2.101) will now read

$$\left.\begin{aligned} &+ \sum_{\nu=0}^{s-1} \beta_\nu z^{\nu+n_\varkappa-s} \underline{p}_\varkappa && \text{for } s = 1,\dots n_\varkappa-1 \\ &- \sum_{\nu=s}^{m} \beta_\nu z^{\nu+n_\varkappa-s} \underline{p}_\varkappa && \text{for } s = n_\varkappa,\dots m \end{aligned}\right\} \quad m \geq n_\varkappa$$

$$- \beta_{s-1} \sum_{\mu=0}^{m-1} \beta_\mu z^\mu \underline{p}_\varkappa + \sum_{\mu=0}^{s-2} \beta_\mu z^{\mu+n_\varkappa-s} \underline{p}_\varkappa \qquad m = n_\varkappa - 1 \tag{2.103}$$

wherein only the range for the index s will have shifted.

In the case $m = n_\varkappa - 1$, $k_{n_\varkappa}^{*(\varkappa)}$ in Eq.(2.102) is determined by means of the following (cf. Eq.(2.98)):

$$k_{n_\varkappa}^{*(\varkappa)} = - \beta_o k_1^{(\varkappa)} - \beta_1 k_2^{(\varkappa)} - \dots - \beta_{m-1} k_{n_\varkappa-1}^{(\varkappa)} \quad .$$

The coefficients $\underline{1}_s^{(\varkappa)}(z)$, $s = 1\dots n_\varkappa-1$ in Eq.(2.101) will then be transformed into

$$\underline{1}_s^{(\varkappa)}(z) \longrightarrow \underline{1}_s^{(\varkappa)}(z) + \beta_{s-1} \underline{p}_\varkappa \Delta_B(z) \ , \qquad s = 1\dots n_\varkappa - 1 \quad .$$

The addends $\beta_{s-1}\underline{p}_\varkappa \Delta_B(z)$ can be added to the final row of (2.101) and (2.103). The row will then read

$$\beta_{s-1}\underline{p}_\varkappa \Delta_B(z) - \beta_{s-1} \sum_{\mu=0}^{m-1} \beta_\mu z^\mu \underline{p}_\varkappa + \sum_{\mu=0}^{s-2} \beta_\mu z^{\mu+n_\varkappa-s} \underline{p}_\varkappa$$

and can be simplified to

$$\beta_{s-1} z^m \underline{p}_\varkappa + \sum_{\mu=0}^{s-2} \beta_\mu z^{\mu+n_\varkappa-s} \underline{p}_\varkappa$$

$$= \sum_{\mu=0}^{s-1} \beta_\mu z^{\mu+n_\varkappa-s} \underline{p}_\varkappa \,, \qquad s = 1,\dots n_\varkappa-1 \,.$$

The above expression has the same appearance as the first row of (2.103) which will therefore now be applicable to $m = n_\varkappa-1$ also.

The acceptance of the second row of Eq.(2.102) onto the left-hand side of Eq.(2.101) will thus cause the last three rows of $\underline{1}_s^{(\varkappa)}(z)$ in Eq.(2.101) to be transformed into

$$\begin{aligned} &+ \sum_{\nu=0}^{s-1} \beta_\nu z^{\nu+n_\varkappa-s} \underline{p}_\varkappa \qquad \text{for } s = 1,\dots n_\varkappa-1 \\ &- \sum_{\nu=s}^{m} \beta_\nu z^{\nu+n_\varkappa-s} \underline{p}_\varkappa \qquad \text{for } s = n_\varkappa,\dots m \;. \end{aligned} \tag{2.104}$$

The second row of (2.104) can be dispensed with in the case $m = n_\varkappa-1$, this being readable from the s range.

After addition also of the first line of Eq.(2.102) we shall finally obtain the following expression for $\underline{z}_i'(z)$ in accordance with Eq.(2.92):

$$\underline{z}_i'(z) = \left[k_{n_1}^{(1)},\ k_{n_2}^{(2)},\ \dots,\ k_{n_p}^{(p)} \right] \underline{P}^{-1} \Delta_B(z) + \tag{2.105}$$

$$+ \underline{1}_1^{'(1)}(z) k_1^{(1)} + \dots + \underline{1}_{n_1-1}^{'(1)}(z) k_{n_1-1}^{(1)} + \underline{1}_{n_1}^{'(1)}(z) k_{n_1}^{*(1)} + \dots + \underline{1}_m^{'(1)}(z) k_m^{*(1)}$$

$$\vdots$$

$$+ \underline{1}_1^{'(p)}(z) k_1^{(p)} + \dots + \underline{1}_{n_p-1}^{'(p)}(z) k_{n_p-1}^{(p)} + \underline{1}_{n_p}^{'(p)}(z) k_{n_p}^{*(p)} + \dots + \underline{1}_m^{'(p)}(z) k_m^{*(p)}$$

with

$$\underline{1}_s^{(\varkappa)}(z) = \sum_{\mu=1}^{\bar{s}} \sum_{\nu=s}^{m} \underline{a}_\mu^{(\varkappa)} \beta_\nu z^{\mu+\nu-s-1} - \sum_{\mu=o}^{s-1} \sum_{\nu=s+1}^{n_\varkappa} \beta_\mu \underline{a}_\nu^{(\varkappa)} z^{\mu+\nu-s-1} + \qquad (2.105a)$$

$$+ \begin{cases} + \sum_{\nu=o}^{s-1} \beta_\nu z^{\nu+n_\varkappa-s} \underline{p}_\varkappa & \text{for } s = 1, \dots n_\varkappa - 1 \\ - \sum_{\nu=s}^{m} \beta_\nu z^{\nu+n_\varkappa-s} \underline{p}_\varkappa & \text{for } s = n_\varkappa, \dots m \end{cases}$$

$\bar{s} = \min(s, n_\varkappa) \qquad s = 1, \dots m \qquad \varkappa = 1, \dots p \qquad m \geq n_o - 1$

$\underline{a}_\mu^{(\varkappa)}$, $\underline{p}_\varkappa$ in accordance with Eqs.(2.99), (2.100).

Eq.(2.105) can be combined further and the following final representation will be obtained for the vector $\underline{z}_i'(z)$ which belongs to the i-th row of the controller transfer function $\underline{Z}_C(z)$ in accordance with Eq.(2.89):

$$\underline{z}_i(z) = \underline{v}(z) + \underline{L}^*(z)\underline{k}^* \qquad (2.106)$$

with

$$\underline{v}(z) = \underline{L}(z)\underline{k} + \Delta_B(z) \sum_{\varkappa=1}^{p} k_{n_\varkappa}^{(\varkappa)} \underline{p}_\varkappa$$

$$\underline{L}(z) = \left[\underline{1}_1^{(1)}(z) \dots \underline{1}_{n_1-1}^{(1)}(z) \,\middle|\, \dots \,\middle|\, \underline{1}_1^{(p)}(z) \dots \underline{1}_{n_p-1}^{(p)}(z)\right]$$

$$\underline{k}' = \left[k_1^{(1)} \dots k_{n_1-1}^{(1)} \,\middle|\, \dots \,\middle|\, k_1^{(p)} \dots k_{n_p-1}^{(p)}\right]$$

$$\underline{L}^*(z) = \left[\underline{1}_{n_1}^{(1)}(z) \dots \underline{1}_m^{(1)}(z) \,\middle|\, \dots \,\middle|\, \underline{1}_{n_p}^{(p)}(z) \dots \underline{1}_m^{(p)}(z)\right]$$

$$\underline{k}'^* = \left[k_{n_1}^{*(1)} \dots k_m^{*(1)} \,\middle|\, \dots \,\middle|\, k_{n_p}^{*(p)} \dots k_m^{*(p)}\right]$$

wherein Eq.(2.105a) is used for $\underline{L}(z)$ and $\underline{L}^*(z)$.

Comprised within the vector $\underline{k}^*$ are all those free parameters which are still available in the i-th partial controller (along with the eigenvalues) and influence the i-th row of the

controller transfer function $\underline{Z}_C(z)$ (and also the i-th row of $\underline{\Gamma}_C(z)$, see section 2.4.2).
Eq.(2.106) is a linear expression in $\underline{k}^*$.

2.4.2 PREPARATION OF $\underline{\Gamma}_C(z)$ FOR THE SENSITIVITY DESIGN

Using Eq.(2.88), the controller transfer function $\underline{\Gamma}_C(z)$ in accordance with Eq.(2.47) will read as follows:

$$\underline{\Gamma}_C(z) = \begin{bmatrix} \underline{i}_1' + \underline{k}_m'^{(1)} \dfrac{\mathrm{adj}(\underline{I}\,z - \underline{F}^{(1)})}{\det(\underline{I}\,z - \underline{F}^{(1)})} (\underline{S}_u^{(1)} - \underline{S}_y^{(1)}\underline{D}) - \underline{k}_y'^{(1)}\underline{D} \\ \vdots \\ \underline{i}_r' + \underline{k}_m'^{(r)} \dfrac{\mathrm{adj}(\underline{I}\,z - \underline{F}^{(r)})}{\det(\underline{I}\,z - \underline{F}^{(r)})} (\underline{S}_u^{(r)} - \underline{S}_y^{(r)}\underline{D}) - \underline{k}_y'^{(r)}\underline{D} \end{bmatrix}.$$

Thus, for $\underline{\Gamma}_C(z)$, we shall have the expression

$$\underline{\Gamma}_C(z) = \begin{bmatrix} \dfrac{\underline{\gamma}_1'(z)}{\Delta_B^{(1)}(z)} \\ \vdots \\ \dfrac{\underline{\gamma}_r'(z)}{\Delta_B^{(r)}(z)} \end{bmatrix} \tag{2.107}$$

with

$$\underline{\gamma}_i'(z) = (\underline{i}_i' - \underline{k}_y'^{(i)}\underline{D})\,\Delta_B^{(i)}(z) + \underline{k}_m'^{(i)}\mathrm{adj}(\underline{I}\,z - \underline{F}^{(i)})(\underline{S}_u^{(i)} - \underline{S}_y^{(i)}\underline{D}) \tag{2.108}$$

and $\Delta_B^{(i)}(z)$ in accordance with Eq.(2.91), wherein i = 1,...r.

For the sake of improved clarity, the index (i) will henceforth again be omitted in the case of the controller matrices and the characteristic polynomials $\Delta_B^{(i)}(z)$.
Using Eq.(2.19a) and Eq.(2.93), the following is applicable for $\underline{S}_u - \underline{S}_y\underline{D}$ in Eq.(2.108):

$$\underline{S}_u - \underline{S}_y\underline{D} = \tilde{\underline{Q}}\,(\tilde{\underline{H}} - [\underline{a}_1 \ldots \underline{a}_p]\underline{P}^{-1}\underline{D}) +$$

$$+ \left[\underline{F}^{n_1}\underline{q}_1 \ldots \underline{F}^{n_p}\underline{q}_p\right]\underline{P}^{-1}\underline{D} \quad ,$$

and for $\underline{k}_y^{'}\underline{D}$, using Eq.(2.27):

$$\underline{k}_y^{'}\underline{D} = [\ldots]\ \underline{P}^{-1}\underline{D} \quad .$$

If we take these relationships into consideration when comparing Eq.(2.108) with Eq.(2.90), then it will become clear that $\underline{\gamma}_i^{'}(z)$ is derived from $\underline{z}_i^{'}(z)$ if, in Eq.(2.93), $[\underline{a}_1 \ldots \underline{a}_p]\underline{P}^{-1}$ is replaced by $\tilde{\underline{H}} - [\underline{a}_1 \ldots \underline{a}_p]\underline{P}^{-1}\underline{D}$ and, on the right-hand side of Eq.(2.93) and in $\underline{k}_y^{'}$, the matrix $\underline{P}^{-1}$ is replaced by $-\underline{P}^{-1}\underline{D}$ and if, moreover, $\underline{i}_i^{'}\Delta_B(z)$ is added to $\underline{z}_i^{'}(z)$.

If, therefore, we here replace the definitions (2.99) and (2.100) by

$$\tilde{\underline{H}} - [\underline{a}_1 \ldots \underline{a}_p]\underline{P}^{-1}\underline{D} =: \begin{bmatrix} \hat{\underline{a}}_1^{'(1)} \\ \vdots \\ \hat{\underline{a}}_{n_1}^{'(1)} \\ \hline \vdots \\ \hline \hat{\underline{a}}_1^{'(p)} \\ \vdots \\ \hat{\underline{a}}_{n_p}^{'(p)} \end{bmatrix} \qquad (2.109)$$

and

$$-\,\underline{P}^{-1}\underline{D} =: \begin{bmatrix} \hat{\underline{p}}_1^{'} \\ \hat{\underline{p}}_2^{'} \\ \vdots \\ \hat{\underline{p}}_p^{'} \end{bmatrix} , \qquad (2.110)$$

then, for $\underline{\gamma}_i(z)$, we shall have the following designation corresponding to Eq.(2.106):

$$\underline{\gamma}_i(z) = \underline{w}(z) + \underline{M}^*(z)\underline{k}^* \qquad (2.111)$$

with

$$\underline{w}(z) = \underline{M}(z)\underline{k} + (\underline{i}_i + \sum_{\varkappa=1}^{p} k_{n_\varkappa}^{(\varkappa)} \hat{\underline{p}}_\varkappa)\, \Delta_B(z)$$

$$\underline{M}(z) = \left[\hat{\underline{1}}_1^{(1)}(z) \; \ldots \; \hat{\underline{1}}_{n_1-1}^{(1)}(z) \,\Big|\, \ldots \,\Big|\, \hat{\underline{1}}_1^{(p)}(z) \ldots \hat{\underline{1}}_{n_p-1}^{(p)}(z) \right]$$

$$\underline{k}' = \left[k_1^{(1)} \; \ldots \; k_{n_1-1}^{(1)} \,\Big|\, \ldots \,\Big|\, k_1^{(p)} \; \ldots \; k_{n_p-1}^{(p)} \right]$$

$$\underline{M}^*(z) = \left[\hat{\underline{1}}_{n_1}^{(1)}(z) \; \ldots \; \hat{\underline{1}}_m^{(1)}(z) \,\Big|\, \ldots \,\Big|\, \hat{\underline{1}}_{n_p}^{(p)}(z) \; \ldots \; \hat{\underline{1}}_m^{(p)}(z) \right]$$

$$\underline{k}'^* = \left[k_{n_1}^{*(1)} \; \ldots \; k_m^{*(1)} \,\Big|\, \ldots \,\Big|\, k_{n_p}^{*(p)} \; \ldots \; k_m^{*(p)} \right]$$

wherein

$$\hat{\underline{1}}_s^{(\varkappa)}(z) = \sum_{\mu=1}^{\bar{s}} \sum_{\nu=s}^{m} \hat{\underline{a}}_\mu^{(\varkappa)} \beta_\nu z^{\mu+\nu-s-1} - \sum_{\mu=o}^{s-1} \sum_{\nu=s+1}^{n_\varkappa} \beta_\mu \hat{\underline{a}}_\nu^{(\varkappa)} z^{\mu+\nu-s-1} +$$

$$+ \begin{cases} + \sum_{\nu=o}^{s-1} \beta_\nu z^{\nu+n_\varkappa-s} \hat{\underline{p}}_\varkappa & \text{for } s = 1,\ldots n_\varkappa-1 \\ - \sum_{\nu=s}^{m} \beta_\nu z^{\nu+n_\varkappa-s} \hat{\underline{p}}_\varkappa & \text{for } s = n_\varkappa,\ldots m \end{cases}$$

$$\bar{s} = \min(s, n_\varkappa) \qquad s = 1,\ldots m \qquad \varkappa = 1,\ldots p \qquad m \geq n_o - 1$$

$\hat{\underline{a}}_\mu^{(\varkappa)}$, $\hat{\underline{p}}_\varkappa$ in accordance with Eq.(2.109), (2.110).

The vector $\underline{k}^*$ occurred previously in Eq.(2.106) and contains all those free parameters which are still available in the i-th partial controller (along with the eigenvalues). Eqs.(2.106) and (2.111) describe the linear influence of $\underline{k}^*$ on the i-th (and only the i-th) row of the controller transfer functions $\underline{Z}_C(z)$ and $\underline{r}_C(z)$ according to Eqs.(2.89) and (2.107) respectively.

In the case of plants without a distribution matrix ($\underline{D}=\underline{O}$), the calculation of $\hat{\underline{1}}_s^{(\varkappa)}(z)$ in Eq.(2.111) becomes simplified because then, in accordance with Eq.(2.110), the $\hat{\underline{p}}$ will disappear and the $\hat{\underline{a}}_\mu^{\prime(\varkappa)}$ in Eq.(2.109) will be transformed into the row vectors of the transformed plant input matrix $\tilde{\underline{H}}$.

2.5 DETERMINATION OF THE FREE PARAMETERS

Free parameters are considered to be those controller parameters which do not have any influence upon the nominal command behavior. All of those free parameters which are available along with the eigenvalues for the i-th partial controller are combined into the vector $\underline{k}^*$ in Eq.(2.106) and in Eq.(2.111). They shall now be utilized for the purpose of satisfying the requirements regarding sensitivity behavior and disturbance behavior indicated in section 2.3. Here, the i-th partial controller will have an influence only on the i-th row of the controller transfer functions $\underline{Z}_C(z)$ and $\underline{r}_C(z)$ in accordance with Eqs.(2.89) and (2.107).

With the minimum partial controller order $m = n_o-1$, free parameters of precisely

$$\eta_o = p\, n_o - n \qquad \text{for } m = n_o-1 \qquad (2.112)$$

will be obtained for a partial controller in accordance with Eq.(2.38) (see Eq.(2.34) in respect of n_o). Where all structural indices of the canonical state space model (2.20), (2.21) are of equal magnitude ($n_1 = n_2 = \ldots = n_p$), the observability index n_o will have its smallest possible value

$n_o = n/p$ and $\eta_o = 0$ will be applicable. Therefore, with the minimum partial controller order $m = n_o-1$, free parameters will only be present where the structural indices of the canonical plant state space model are not all of equal magnitude. The maximum quantity of free parameters will be obtained where only one of the p structural indices is of a magnitude greater than one. The observability index n_o will then have the greatest possible value, this being $n_o = n-p+1$, and the quantity of free parameters in accordance with Eq.(2.112) will be $\eta_o = (n-p)(p-1)$.

It can be further deduced from Eq.(2.38) that each time the order of the partial controller is increased by the value of one, additional free parameters, depending precisely upon p, will be furnished for the partial controller in question.

In normal cases, we shall select the same order and the same eigenvalues for all partial controllers. Following this, the free parameters comprised within $\underline{k}^*$ will be determined for each partial controller, as described in sections 2.5.1 , 2.5.2 or 2.5.3 .

2.5.1 ZEROS FOR THE SENSITIVITY FUNCTION

If, with the frequencies

$$\varphi = \varphi_\mu , \qquad \mu = 1 \dots s \tag{2.113}$$

the sensitivity function $\underline{S}(e^{j\varphi_\mu})$ (or, in the case $p > r$, only r rows of $\underline{S}(e^{j\varphi_\mu})$) should disappear, the the requirement (2.68) will be the decisive factor here in the case $p = r$. Upon utilizing the representation (2.111) for the i-th row of $\underline{r}_C(z)$, cf. Eq.(2.107), we shall have the following requirements for the i-th partial controller:

$$\underline{\gamma}_i(e^{j\varphi_\mu}) = \underline{w}(e^{j\varphi_\mu}) + \underline{M}^*(e^{j\varphi_\mu})\underline{k}^* = \underline{0} , \tag{2.114a}$$

$$\mu = 1 \dots s .$$

(In the case $p < r$, the requirement (2.68) will relate only to p columns of $\underline{\Gamma}_C$, cf. section 2.3.1.3, so that, in this case, the condition (2.114a) will need to be satisfied only for p elements of $\underline{\gamma}_i$).

In the case $p > r$, the requirement (2.79) will also apply in addition to requirement (2.68). Since requirement (2.79) relates only to (p-r) columns of $\underline{Z}_C(z)$, only the corresponding (p-r) elements of the vectors $\underline{z}_i'$ in Eq.(2.89) need to be observed. Henceforth, this will be observed by means of a line over $\underline{z}_i$. In the same way, in the representation (2.106) for $\underline{z}_i$, a line placed over $\underline{v}(z)$ and $\underline{L}^*(z)$ will indicate that only the corresponding (p-r) elements of $\underline{v}(z)$ or (p-r) rows of $\underline{L}^*(z)$ need to be considered. In the case $p > r$, therefore, (2.114a) will be completed by the condition

$$\bar{\underline{z}}_i(e^{j\varphi_\mu}) = \bar{\underline{v}}(e^{j\varphi_\mu}) + \bar{\underline{L}}^*(e^{j\varphi_\mu})\underline{k}^* = \underline{0} \quad , \qquad (2.114b)$$

$$\mu = 1 \ldots s \; .$$

Note :

If an λ-fold zero $z_\mu = e^{j\varphi_\mu}$ is involved, then along with $\underline{\gamma}_i(z)$ and $\bar{\underline{z}}_i(z)$, even the differential quotients of these functions as far as the (λ-1)-th derivation must also disappear. The derivations of $\underline{\gamma}_i(z)$ and $\underline{z}_i(z)$ after z can be formed easily since z occurs only in power form; see Eqs.(2.106) and (2.111).

If we break the conditions (2.114a,b) into real and imaginary components, then we shall obtain the following linear set of equations combined for $\underline{k}^*$:

$$\underline{1}_o + \underline{L}_o^* \, \underline{k}^* = \underline{0} \qquad (2.115)$$

with

$$\underline{l}_o = \begin{bmatrix} Re\{\underline{w}(z_1)\} \\ Im\{\underline{w}(z_1)\} \\ Re\{\bar{\underline{v}}(z_1)\} \\ Im\{\bar{\underline{v}}(z_1)\} \\ Re\{\underline{w}(z_2)\} \\ Im\{\underline{w}(z_2)\} \\ \vdots \\ Re\{\bar{\underline{v}}(z_s)\} \\ Im\{\bar{\underline{v}}(z_s)\} \end{bmatrix} \quad \text{und} \quad \underline{\underline{L}}_o^* = \begin{bmatrix} Re\{\underline{M}^*(z_1)\} \\ Im\{\underline{M}^*(z_1)\} \\ Re\{\bar{\underline{L}}^*(z_1)\} \\ Im\{\bar{\underline{L}}^*(z_1)\} \\ Re\{\underline{M}^*(z_2)\} \\ Im\{\underline{M}^*(z_2)\} \\ \vdots \\ Re\{\bar{\underline{L}}^*(z_s)\} \\ Im\{\bar{\underline{L}}^*(z_s)\} \end{bmatrix} ,$$

wherein $z_\mu = e^{j\varphi_\mu}$.

It can be seen from the above observations that in all three cases $p > r$, $p = r$ and $p < r$ the two vectors (2.114a) and (2.114b) together possess (complex) elements in precise keeping with p. Each φ_μ of (2.113) thus contributes 2p rows to the set of equations, so that Eq.(2.115) will contain a total of 2ps rows. The sole exceptions here are the frequencies $\varphi_\mu = 0$ and $\varphi_\mu = \pi$ because with these the imaginary components will disappear and only the relevant contribution of p rows to the set of equations (2.115) will take place.

In order to enable the set of equations (2.115) to have a solution, $\underline{L}_o^*$ must be regular with regard to rows, i.e. the number of elements in $\underline{k}^*$ must (at least) correspond to the number of rows of $\underline{L}_o^*$.

Since, in the case of a partial controller of the minimum order $m = n_o - 1$, the quantity of free parameters according to Eq.(2.112) comprised within $\underline{\underline{k}}^*$ is not normally adequate for the above, it will be necessary for the order of the partial controllers to be correspondingly increased. In this respect, compare what has been said with regard to Eq.(2.112).

Attention should again be given to section 2.2.7.2.3, whereby all frequencies (2.113) must differ from the plant zeros.

2.5.2 MINIMIZATION OF THE SENSITIVITY FUNCTION

In the case p = r, the requirement (2.65) for minimization of the sensitivity function in the frequency range furnished by the frequencies

$$\varphi = \varphi_\nu \quad , \qquad \nu = 1 \dots q \tag{2.116}$$

has led to the requirement (2.71),

$$\sum_{\nu=1}^{q} \alpha^2(\varphi_\nu)\, b^2(\varphi_\nu) \left\| \underline{r}_C(e^{j\varphi_\nu}) \right\|^2 \longrightarrow \text{as small as possible} \tag{2.117a}$$

with

$$b(\varphi_\nu) = \left\| \frac{\underline{r}_R(e^{j\varphi_\nu})}{\Delta_R(e^{j\varphi_\nu})} \right\| ,$$

which, in the case p > r, has been completed by the condition (2.83)

$$\sum_{\nu=1}^{q} \alpha^2(\varphi_\nu)c^2(\varphi_\nu) \left\| \underline{Z}_{C2}(e^{j\varphi_\nu}) \right\|^2 \longrightarrow \text{as small as possible} \tag{2.117b}$$

with

$$c(\varphi_\nu) = \left\| \underline{G}_1(e^{j\varphi_\nu}) \frac{\underline{r}_R(e^{j\varphi_\nu})}{\Delta_R(e^{j\varphi_\nu})} \right\| .$$

(In the case p < r, as in the case p = r, we merely have the condition (2.117a) which, however, will then relate only to p columns of $\underline{r}_C$, cf. section 2.3.1.3).

It will be practical to combine (2.117a) and (2.117b) and this will provide

$$\sum_{\nu=1}^{q} \alpha^2(\varphi_\nu)b^2(\varphi_\nu) \left\| \underline{r}_C(e^{j\varphi_\nu}) \right\|^2 + \alpha^2(\varphi_\nu)c^2(\varphi_\nu) \left\| \underline{Z}_{C2}(e^{j\varphi_\nu}) \right\|^2 \longrightarrow \text{as small as possible} \tag{2.118}$$

Upon employing the Euclidean matrix norm, the square of norm of a matrix will equal the sum of the magnitude squares of the row vectors. Taking account of Eq.(2.89) and Eq.(2.107), we shall then obtain the following requirement from (2.118) for the i-th partial controller:

$$\sum_{\nu=1}^{q} \tilde{b}^2(\varphi_\nu)|\underline{\gamma}_i(e^{j\varphi_\nu})|^2 + \tilde{c}^2(\varphi_\nu)|\bar{\underline{z}}_i(e^{j\varphi_\nu})|^2$$

$$\longrightarrow \text{ as small as possible} \quad (2.119)$$

$$\text{with } \tilde{b}(\varphi_\nu) = \alpha(\varphi_\nu)\frac{b(\varphi_\nu)}{|\Delta_B(e^{j\varphi_\nu})|} \quad , \quad \tilde{c}(\varphi_\nu) = \alpha(\varphi_\nu)\frac{c(\varphi_\nu)}{|\Delta_B(e^{j\varphi_\nu})|} ,$$

wherein $\bar{\underline{z}}_i$ contains only (p-r) elements of $\underline{z}_i$, cf. section 2.5.1 . The expression (2.119) is the magnitude square of the following vector:

$$\begin{bmatrix} \tilde{b}(\varphi_1)\ \mathrm{Re}\{\underline{\gamma}_i(e^{j\varphi_1})\} \\ \tilde{b}(\varphi_1)\ \mathrm{Im}\{\underline{\gamma}_i(e^{j\varphi_1})\} \\ \tilde{c}(\varphi_1)\ \mathrm{Re}\{\bar{\underline{z}}_i(e^{j\varphi_1})\} \\ \tilde{c}(\varphi_1)\ \mathrm{Im}\{\bar{\underline{z}}_i(e^{j\varphi_1})\} \\ \tilde{b}(\varphi_2)\ \mathrm{Re}\{\underline{\gamma}_i(e^{j\varphi_2})\} \\ \vdots \\ \tilde{c}(\varphi_q)\ \mathrm{Im}\{\bar{\underline{z}}_i(e^{j\varphi_q})\} \end{bmatrix} .$$

If Eq.(2.106) and Eq.(2.111) are taken into consideration, then (2.119) will be transformed into the requirement

$$|\ \underline{l} + \underline{L}^* \underline{k}^*\ |^2 \longrightarrow \text{ as small as possible} \qquad (2.120)$$

with

$$\underline{l} = \begin{bmatrix} \tilde{b}(\varphi_1)\ \mathrm{Re}\{\underline{w}(z_1)\} \\ \tilde{b}(\varphi_1)\ \mathrm{Im}\{\underline{w}(z_1)\} \\ \tilde{c}(\varphi_1)\ \mathrm{Re}\{\underline{\bar{v}}(z_1)\} \\ \tilde{c}(\varphi_1)\ \mathrm{Im}\{\underline{\bar{v}}(z_1)\} \\ \tilde{b}(\varphi_2)\ \mathrm{Re}\{\underline{w}(z_2)\} \\ \cdot \\ \cdot \\ \cdot \\ \tilde{c}(\varphi_q)\ \mathrm{Im}\{\underline{\bar{v}}(z_q)\} \end{bmatrix} \quad \text{and} \quad \underline{L}^* = \begin{bmatrix} \tilde{b}(\varphi_1)\ \mathrm{Re}\{\underline{M}^*(z_1)\} \\ \tilde{b}(\varphi_1)\ \mathrm{Im}\{\underline{M}^*(z_1)\} \\ \tilde{c}(\varphi_1)\ \mathrm{Re}\{\underline{\bar{L}}^*(z_1)\} \\ \tilde{c}(\varphi_1)\ \mathrm{Im}\{\underline{\bar{L}}^*(z_1)\} \\ \tilde{b}(\varphi_2)\ \mathrm{Re}\{\underline{M}^*(z_2)\} \\ \cdot \\ \cdot \\ \cdot \\ \tilde{c}(\varphi_q)\ \mathrm{Im}\{\underline{\bar{L}}^*(z_q)\} \end{bmatrix},$$

wherein $z = e^{j\varphi_\nu}$.

The magnitude square (2.120) will assume its smallest value where $\underline{k}^*$ is the solution vector of the linear set of equations

$$\underline{L}'^* \underline{l} + \underline{L}'^* \underline{L}^* \underline{k}^* = \underline{0} \quad . \tag{2.121}$$

The expression (2.118) will therefore be minimized in relation to the free parameters present in the controller along with the eigenvalues if $\underline{k}^*$ is determined in accordance with Eq.(2.121) for each partial controller.

The order of the set of equations (2.121) is independent of the quantity q of frequency points (2.116) which are placed above the frequency range in question. The calculation, however, for formation of $\underline{l}$, $\underline{L}^*$ and $\underline{L}'^*\underline{L}^*$ will become more burdensome with an increasing quantity of frequency points, so these should be accordingly limited to a practical number. The effect of minimization depends not so much on a greatest possible quantity of frequency points but much more on the dimension of the vector $\underline{k}^*$. The greater the order of the partial controllers and, accordingly, the quantity of free parameters, then the smaller the values accepted by the norm of the sensitivity matrix in the frequency range furnished by the frequency points (2.116).

Even so, the quantity of frequency points must at least be large enough to enable $\underline{L}^*$ in Eq.(2.121) to remain regular as to columns, though not as to rows. Where $\underline{L}^*$ is regular as to rows (i.e. where $\underline{L}'^*$ is regular as to columns), Eq.(2.121) would, in fact, become transformed into the resolvable set of equations $\underline{1} + \underline{L}^*\underline{k}^* = \underline{0}$ whereby, with the frequencies(2.121), zeros would be generated for the sensitivity function; cf. section 2.5.1. A practical selection of the frequency points is likely to be found in the five- to ten-fold of this lower limiting value.

2.5.3 MINIMIZATION OF THE SENSITIVITY FUNCTION WITH SECONDARY CONDITION IN ACCORDANCE WITH 2.5.1

If the sensitivity function at the frequencies (2.116) is to be minimized in accordance with the requirement (2.65) (section 2.5.2) and, in addition, is to possess zeros at the frequencies (2.113) (section 2.5.1), then the following problem will arise:

> It will be necessary to find a $\underline{k}^*$ in order to enable Eq.(2.115) to be satisfied and the norm-square (2.120) to assume its smallest value.

The following theorem from the non-linear programming [13] is of significance in this respect:

The following will be applicable

$$f(\underline{x}^\circ) = \min_{\underline{x}\in\mathbb{R}^\alpha} \{f(\underline{x}) \mid \underline{g}(\underline{x}) = \underline{0}\}$$

$$\text{with} \quad f: \mathbb{R}^\alpha \rightarrow \mathbb{R}^1 \quad \text{convex}$$

$$\underline{g} = \begin{bmatrix} g_1 \\ \vdots \\ g_t \end{bmatrix} : \mathbb{R}^\alpha \rightarrow \mathbb{R}^t \quad \text{affine},$$

if and only if $\underline{x}_o$ with $\underline{g}(\underline{x}°) = \underline{0}$ is also able to satisfy the condition

$$\mathrm{grad}\{f(\underline{x}°)\} + \sum_{i=1}^{t} \lambda_i \, \mathrm{grad}\{g_i(\underline{x}°)\} = \underline{0} \qquad (2.122)$$

for a $\underline{\lambda}' := [\lambda_1 \ \ldots \ \lambda_t]$.

With $\underline{g}(\underline{k}^*) := \underline{1}_o + \underline{L}_o^* \underline{k}^*$ in accordance with Eq.(2.115) and the convex function $f(\underline{k}^*) := |\underline{1} + \underline{L}^* \underline{k}^*|^2$ in accordance with Eq.(2.120), Eq.(2.122) will be transformed into

$$2\,\underline{L}'^*(\underline{1} + \underline{L}^* \underline{k}^*) + \underline{L}_o'^* \underline{\lambda} = \underline{0} \ . \qquad (2.123)$$

The combination of Eq.(2.115) and Eq.(2.123) will finally provide

$$\begin{matrix} \eta \\ 2ps \end{matrix} \left\{ \begin{bmatrix} \overbrace{2\,\underline{L}'^* \underline{L}^*}^{\eta} & \overbrace{\underline{L}_o'^*}^{2ps} \\ \underline{L}_o^* & \underline{0} \end{bmatrix} \begin{bmatrix} \underline{k}^* \\ \underline{\lambda} \end{bmatrix} = - \begin{bmatrix} 2\,\underline{L}'^* \underline{1} \\ \underline{1}_o \end{bmatrix} \right. \qquad (2.124)$$

The numerical value obtained for the vector $\underline{\lambda}$ has no significance for the dimensioning of the controller.

Taking into account the secondary condition (2.115), the normsquare (2.120) will thus assume its smallest value (as a function of $\underline{k}^*$) if $\underline{k}^*$ is determined according to the set of equations (2.124).

The accordingly necessary increasing of the partial controller order over and above the minimum value n_o-1 will correspond to the sum of the increases which would be needed for resolution of the individual problems in accordance with sections 2.5.1 and 2.5.2 .

2.6 SUMMARY OF THE CONTROLLER DESIGN

The designing of the controller will be effected in six steps: *)

I. Ascertaining of the discrete model (2.1) for the plant.

II. Determination of the state gain matrix $\underline{K}$ (and of the prefactor $\underline{\rho}$) in accordance with the required command behavior.

III. Transformation of plant and gain matrix into the canonical multivariable form (2.20)/(2.21)

$$\underline{\tilde{\Phi}} = \left[\underline{i}_2 \ldots \underline{i}_{n_1}, \underline{a}_1 \;\vdots\; \underline{i}_{n_1+2} \ldots \underline{i}_{n_1+n_2}, \underline{a}_2 \;\vdots\; \ldots \; \underline{i}_n, \underline{a}_p\right]$$

$\underline{\tilde{H}}$ without particular form

$$\underline{\tilde{C}} = \underline{P}\,\underline{\hat{C}}$$

with

$$\underline{P} = \begin{bmatrix} 1 & & & \\ p_{21} & 1 & & \underline{0} \\ \vdots & & \ddots & \\ p_{p1} & p_{p2} & \cdots & 1 \end{bmatrix} \qquad \underline{\hat{C}} = \begin{bmatrix} \underline{i}'_{n_1} \\ \underline{i}'_{n_1+n_2} \\ \vdots \\ \underline{i}'_n \end{bmatrix}$$

$$\underline{\tilde{k}}' = \left[k_1^{(1)} \ldots k_{n_1}^{(1)} \;\vdots\; k_1^{(2)} \ldots k_{n_2}^{(2)} \;\vdots\; \ldots \;\vdots\; k_1^{(p)} \ldots k_{n_p}^{(p)}\right]$$

(one row of $\underline{\tilde{K}}$ in accordance with Eq.(2.18))

IV. Sensitivity design:
The necessary quantity of free parameters and, accordingly, the necessary partial controller order are obtained through the requirements regarding disturbance behavior and sensitivity behavior (section 2.5). Calculation of

the free parameters ($k_{n_i}^{*(i)} \ldots k_m^{*(i)}$, i = 1...p) of each individual partial controller in accordance with sections 2.5.1, Eq.(2.115), 2.5.2, Eq.(2.121) or 2.5.3, Eq.(2.124).

V. Calculation of the partial controller matrices in accordance with the following scheme:

1.

$$\underline{F} = \begin{bmatrix} 0 & 1 & & \underline{0} \\ & & \ddots & \\ 0 & 0 & & 1 \\ -\beta_o & -\beta_1 & \ldots & -\beta_{m-1} \end{bmatrix}$$

2. $\underline{k}_m' = [1 \quad 0 \ldots \quad 0]$

3. $\underline{S}_u = \tilde{\underline{Q}}\ \tilde{\underline{H}}$ with

$$\tilde{\underline{Q}} = \left[\begin{array}{ccc|c|ccc} k_1^{(1)} \ldots & k_{n_1-1}^{(1)} & k_{n_1}^{*(1)} & & k_1^{(p)} \ldots & k_{n_p-1}^{(p)} & k_{n_p}^{*(p)} \\ k_2^{(1)} \ldots & k_{n_1}^{*(1)} & k_{n_1+1}^{*(1)} & \cdots & k_2^{(p)} \ldots & k_{n_p}^{*(p)} & k_{n_p+1}^{*(p)} \\ & \iddots & \vdots & & & \iddots & \vdots \\ & & k_{n_1+m-1}^{*(1)} & & & & k_{n_p+m-1}^{*(p)} \end{array}\right]$$

4. $\underline{k}_y' = \left[k_{n_1}^{(1)} - k_{n_1}^{*(1)}, \ \ldots, \ k_{n_p}^{(p)} - k_{n_p}^{*(p)}\right] \underline{P}^{-1}$

5.

$$\underline{S}_y = \left\{ \tilde{\underline{Q}}[\underline{a}_1 \ldots \underline{a}_p] - \left[\begin{array}{c|c|c} k_{n_1+1}^{*(1)} & & k_{n_p+1}^{*(p)} \\ k_{n_1+2}^{*(1)} & \cdots & k_{n_p+2}^{*(p)} \\ \vdots & & \vdots \\ k_{n_1+m}^{*(1)} & & k_{n_p+m}^{*(p)} \end{array}\right] \right\} \underline{P}^{-1}$$

The "starred" variables with an index greater than m occurring in 3, 4 and 5 are furnished by means of

$$k_j^{*(\xi)} = - \left[k_{j-m}^{(*)(\xi)}, \ldots, k_{j-1}^{(*)(\xi)}\right] \begin{bmatrix} ß_o \\ ß_1 \\ \vdots \\ ß_{m-1} \end{bmatrix} , \quad j > m ,$$

i.e. by means of the negative scalar product of the variables preceding m in each case with the vector $[ß_o \quad ß_1 \ldots ß_{m-1}]'$.

VI. Check of results by means of computer simulation.

*) Items IV and V only are dealt with in the present work.

2.7 EXAMPLE

In respect of the plant, the following example concerns two electric generators connected to each other via a tie line. The aim of the control is to make the nominal system frequency in each generator and the tie-line loading constant as insensitive as possible to fluctuations in the demand loads.

The linearized state space model of the plant ($n = 7$, $p = r = 2$) reads as follows [14], [9]:

$$\dot{\underline{x}}(t) = \begin{bmatrix} -a_{11} & 0 & a_{13} & -a_{14} & 0 & 0 & 0 \\ -a_{21} & -a_{22} & 0 & 0 & 0 & 0 & 0 \\ 0 & a_{32} & -a_{33} & 0 & 0 & 0 & 0 \\ a_{41} & 0 & 0 & 0 & -a_{45} & 0 & 0 \\ 0 & 0 & 0 & -a_{54} & -a_{55} & 0 & a_{57} \\ 0 & 0 & 0 & 0 & -a_{65} & -a_{66} & 0 \\ 0 & 0 & 0 & 0 & 0 & a_{76} & -a_{77} \end{bmatrix} \underline{x}(t) + \tag{2.125}$$

$$+ \begin{bmatrix} 0 & 0 \\ b_{21} & 0 \\ 0 & 0 \\ 0 & 0 \\ 0 & 0 \\ 0 & b_{62} \\ 0 & 0 \end{bmatrix} \underline{u}(t) + \begin{bmatrix} -e_{11} & 0 \\ 0 & 0 \\ 0 & 0 \\ 0 & 0 \\ 0 & -e_{52} \\ 0 & 0 \\ 0 & 0 \end{bmatrix} \underline{\xi}(t)$$

$$\underline{y}(t) = \begin{bmatrix} 1 & 0 & 0 & 0 & 0 & 0 & 0 \\ 0 & 0 & 0 & 1 & 0 & 0 & 0 \end{bmatrix} \underline{x}(t)$$

with the coefficients

$$a_{11} = \frac{D_1\ f^*}{2H_1} \qquad a_{13} = a_{14} = \frac{f^*}{2H_1} \qquad a_{21} = \frac{1}{T_{gv1}R_1} \qquad a_{22} = \frac{1}{T_{gv1}}$$

$$a_{32} = a_{33} = \frac{1}{T_{t1}} \qquad a_{41} = a_{45} = T_{12}^* \qquad a_{54} = k_{12}\frac{f^*}{2H_2} \qquad a_{55} = \frac{D_2 f^*}{2H_2}$$

$$a_{57} = \frac{f^*}{2H_2} \qquad a_{65} = \frac{1}{T_{gv2}R_2} \qquad a_{66} = \frac{1}{T_{gv2}} \qquad a_{76} = a_{77} = \frac{1}{T_{t2}}$$

$$b_{21} = \frac{1}{T_{gv1}} \qquad b_{62} = \frac{1}{T_{gv2}} \qquad e_{11} = \frac{f^*}{2H_1} \qquad e_{52} = \frac{f^*}{2H_2}$$

and the values:

inertia constant	$H_1 = H_2 = 5$ s	
load frequency constant	$D_1 = D_2 = 8.33\ 10^{-3}$ pu MW/Hz	
nominal system frequency	$f^* = 60$ Hz	
speed governor time constant	$T_{gv1} = T_{gv2} = 0.08$ s	(2.126)
turbine time constant	$T_{t1} = T_{t2} = 0.3$ s	
tie-line power-flow constant	$T_{12}^* = 0.545$ pu MW	
self-regulation of generator	$R_1 = R_2 = 2.4$ Hz/pu MW	
	$k_{12} = 1$.	

The state vector $\underline{x}$, the input vector $\underline{u}$ and the disturbance vector $\underline{\xi}$ are defined as follows:

$$\underline{x} = \begin{bmatrix} \Delta f_1 \\ \Delta X_{gv1} \\ \Delta P_{g1} \\ \Delta P_{tie} \\ \Delta f_2 \\ \Delta X_{gv2} \\ \Delta P_{g2} \end{bmatrix} \qquad \underline{u} = \begin{bmatrix} \Delta P_{c1} \\ \\ \Delta P_{c2} \end{bmatrix} \qquad \underline{\xi} = \begin{bmatrix} \Delta P_{d1} \\ \\ \Delta P_{d2} \end{bmatrix}$$

with the meanings

Δf	incremental frequency deviation
ΔX_{gv}	incremental change in governor valve position
ΔP_g	incremental generation change
ΔP_{tie}	incremental change in tie-line power
ΔP_c	incremental change in speed-changer position
ΔP_d	incremental load demand change .

The following conditions need to be satisfied:

No stationary deviations must occur as a result of a disturbance input jump in the case of the frequency and the tie-line power. During the control procedure, the frequency deviations in accordance with the magnitude must not exceed 0.02 Hz with reference to a disturbance input jump of $\Delta P_{d1} = 0.01$ pu MW.

(The nominal system frequency f_2 always has the same deviation as f_1 and the maximum amplitudes of Δf_2 lie below those of Δf_1. For this reason, only Δf_1 will be considered).

We shall thus have the following requirements for the output vector $\underline{y}' =: [y_1 \quad y_2]$:

$$\lim_{t\to\infty} \underline{y}(t) = \underline{0} \qquad \text{for} \quad \underline{\xi}(t) = \underline{\alpha}\sigma(t) \qquad (2.127)$$

$$(\underline{\alpha} \text{ arbitrary})$$

and

$$- 0.02 \leq y_1(t) \leq 0.02 \qquad (2.128)$$

for a disturbance input jump

$$\underline{\xi}(t) = \begin{bmatrix} 0.01 \\ \\ 0 \end{bmatrix} \sigma(t) \quad . \qquad (2.129)$$

0.2 sec has been selected as the sampled period and the gain matrix $\underline{K}$ determined in accordance with the quadratic index of performance

$$\sum_{\nu=1}^{\infty} \underline{y}_\nu' \underline{y}_\nu + 0.1\, \underline{u}_\nu' \underline{u}_\nu \longrightarrow \min . \qquad (2.130)$$

The transformation of the discrete model into Luenberger's second form will then furnish

$$\begin{aligned} \underline{\tilde{x}}_{\nu+1} &= \underline{\tilde{\Phi}}\, \underline{\tilde{x}}_\nu + \underline{\tilde{H}}\, \underline{u}_\nu \\ \underline{y}_\nu &= \underline{\tilde{C}}\, \underline{\tilde{x}}_\nu \end{aligned} \qquad (2.131)$$

with

$$\underline{\tilde{\Phi}} = \begin{bmatrix} 0 & 0 & 0 & -.25823 & 0 & 0 & .092031 \\ 1 & 0 & 0 & 4.0906 & 0 & 0 & -1.3859 \\ 0 & 1 & 0 & -8.0123 & 0 & 0 & 2.2289 \\ 0 & 0 & 1 & 4.0162 & 0 & 0 & 0 \\ 0 & 0 & 0 & .13601 & 0 & 0 & -.041722 \\ 0 & 0 & 0 & -2.0648 & 1 & 0 & .57677 \\ 0 & 0 & 0 & 3.0029 & 0 & 1 & -.42011 \end{bmatrix}$$

$$\underline{\tilde{C}} = \begin{bmatrix} 0 & 0 & 0 & 1 & 0 & 0 & 0 \\ \\ 0 & 0 & 0 & -1.6294 & 0 & 0 & 1 \end{bmatrix}$$

$$\tilde{\underline{H}} = \begin{bmatrix} -.091431 & .0012867 \\ -.84763 & .020847 \\ -.11463 & .020817 \\ .16294 & .0013355 \\ .057145 & -.00050827 \\ .54761 & -.0063161 \\ .27057 & -.0029081 \end{bmatrix}$$

and

$$\tilde{\underline{K}} = \begin{bmatrix} -.88481 & -.26036 & 3.5390 & 9.8241 & -.51027 & -1.3309 & .88247 \\ 8.9550 & 12.335 & 13.194 & 11.048 & 7.1317 & 11.946 & 14.509 \end{bmatrix}$$

(2.132)

The structural indices will thus have the values

$$n_1 = 4 \ , \quad n_2 = 3$$

and the observability index will have the value

$$n_o = 4 \quad . \tag{2.133}$$

The eigenvalues of the command behavior, i.e. the zeros of $\Delta_R(z) = \det(\underline{I}_n z - \underline{\Phi} + \underline{H}\ \underline{K})$, will read

$$\begin{aligned} z_1 &= 0.0573 \\ z_2 &= 0.0714 \\ z_{3;4} &= 0.143 \pm 0.403j \\ z_{5;6} &= 0.581 \pm 0.497j \\ z_7 &= 0.605 \quad . \end{aligned} \tag{2.134}$$

The prefactor

$$\underline{\rho} = \begin{bmatrix} 1.5693 & .36414 \\ .60959 & -2.6336 \end{bmatrix} \tag{2.135}$$

will be obtained from the stationary requirement at the nominal command behavior, cf. Eq.(2.45),

$$\underline{G}(1)\,\frac{\underline{\Gamma}_R(1)}{\Delta_R(1)}\,\underline{\rho} \stackrel{!}{=} \underline{I}_2 \quad .$$

The frequency spectrum of the discrete plant - Eq.(2.42) with $z = e^{j\varphi}$ - and the required magnitude of control input for the command behavior - Eq.(2.41) with $z = e^{j\varphi}$ - are illustrated in Figs. 8 and 9 . The product of the two variables will furnish the frequency spectrum of command behavior in accordance with Eq.(2.45) and is illustrated in Fig.10.

In the present example, prominence has been given to the disturbance behavior. Nevertheless, the command behavior too will always be taken into consideration henceforth in order to demonstrate the complete feedback system behavior.

Fig.11 shows the feedback system's nominal command behavior, which is not influenced by the subsequent sensitivity design of the partial controllers. The behavior of the plant's output variables has been illustrated for a command input jump of

$$\underline{r}_\nu = \begin{bmatrix} 0.01 \\ 0.01 \end{bmatrix} \quad . \tag{2.136}$$

Henceforth, the nominal disturbance behavior will be examined, along with the disturbance and command behavior under the following set of plant parameters which have become altered in relation to the nominal values (2.126):

$$
\begin{array}{ll}
H_1 = 4 \text{ s} & H_2 = 6 \text{ s} \\
D_1 = 10^{-2} \text{ pu MW/Hz} & D_2 = 9 \cdot 10^{-3} \text{ pu MW/Hz} \\
T_{gv1} = 0.0625 \text{ s} & T_{gv2} = 0.1 \text{ s} \\
T_{t1} = 0.33 \text{ s} & T_{t2} = 0.37 \text{ s} \\
T_{12}^{*} = 0.5 \text{ pu MW} & \\
R_1 = 2.65 \text{ Hz/pu MW} & R_2 = 2 \text{ Hz/pu MW} \\
k_{12} = 1.2 \;. &
\end{array}
\qquad (2.137)
$$

The above will produce modifications of up to 50% in the system parameters of the model (2.125) of the continuous plant.

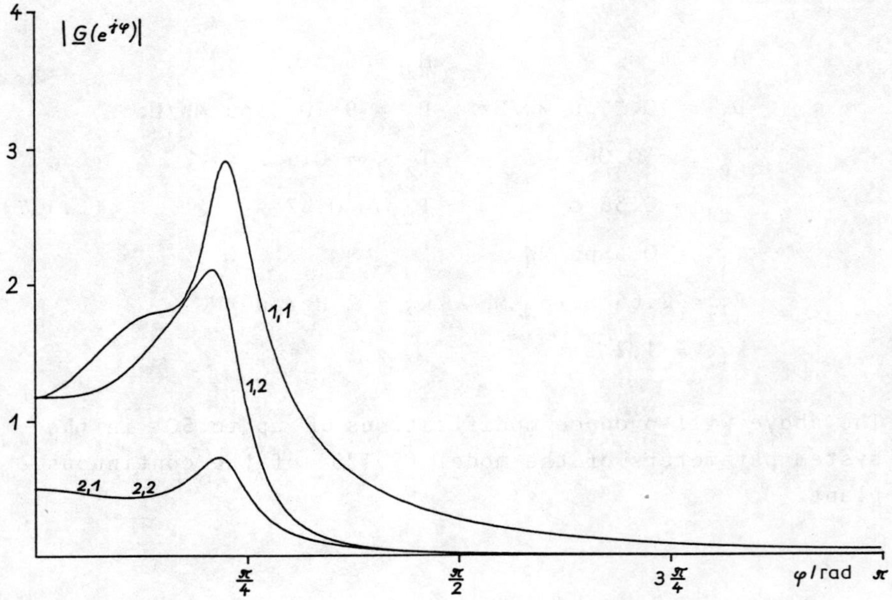

Fig.2.8: Amplitude frequency spectrum of discrete plant (nominal case with parameter set 2.126)

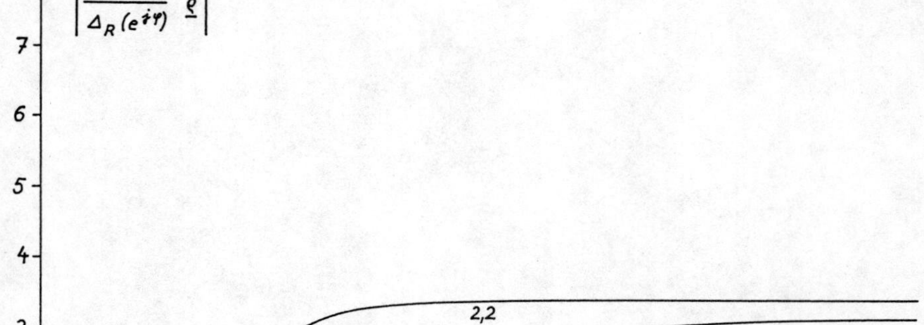

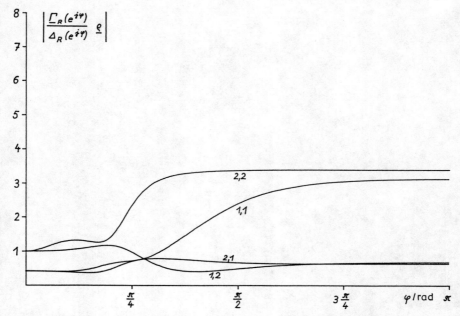

Fig.2.9: Required expense of control input for command behavior (nominal case)

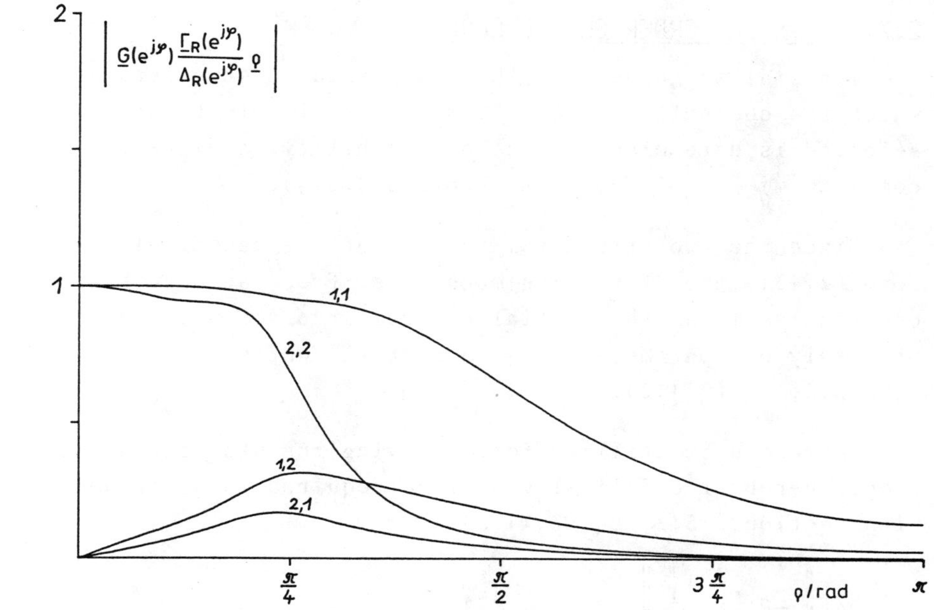

Fig.2.10: Amplitude frequency spectrum of command behavior (nominal case)

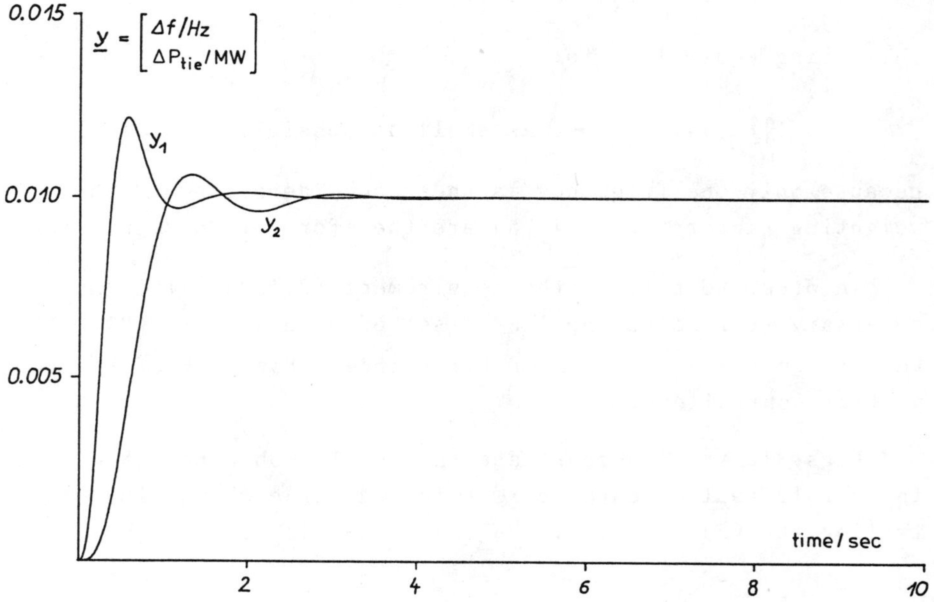

Fig.2.11: command behavior (nominal case)

2.7.1 2 x 3rd ORDER CONTROLLERS

The smallest order for the two partial controllers at which the controller eigenvalues are still able to be freely selected is determined by the observability index (2.133) and comes to $m_o = n_o - 1 = 3$ (cf. section 2.2.5).

Since the two structural indices of the canonical state form (2.131) are of varying magnitude, free parameters will occur as early as the partial controller order m_o, and with precisely one parameter in each partial controller, cf. section 2.5, Eq.(2.112).

These can be utilized for improving the stationary precision, whereby the following will be required in accordance with section 2.5.2, Eq.(2.117a):

$$\alpha^2(\varphi_1) \left\| \frac{\underline{r}_R(e^{j\varphi_1})}{\Delta_R(e^{j\varphi_1})} \right\|^2 \quad ||\underline{r}_C(e^{j\varphi_1})||^2 \longrightarrow \text{ as small as possible}$$

$$\text{with} \quad \varphi_1 = 0 \quad ,$$

this being equivalent to

$$||\underline{r}_C(1)|| \longrightarrow \text{ as small as possible} \quad , \qquad (2.138)$$

because only one frequency is under consideration and the weighting factors in (2.117a) are therefore of no significance.

In order to satisfy the requirement (2.138), it is merely necessary to resolve the linear set of equations (2.121), in the present case a system of first order, for each of the two partial controllers.

Figs. 12 to 15 demonstrate the results obtained when a three-fold real eigenvalue is selected for each partial controller at $\lambda(\underline{F}) = 0.4$, i.e. $\Delta_B(z) = (z-0.4)^3$.

The following illustrations are given:

Fig. 12: the magnitudes of the elements

of $\dfrac{\underline{r}_R(e^{j\varphi})}{\Delta_R(e^{j\varphi})}\ \underline{r}_C(e^{j\varphi})$, cf.Eqs.(2.59) and (269)/(2.70)

Fig. 13: the control variables with a command input jump (2.136) and with the changed set of plant parameters (2.137)

Fig. 14: the control variables with a disturbance input jump (2.129)

Fig. 15: the control variables with a disturbance input jump (2.129) and with the changed set of plant parameters (2.137).

The offsets resulting from disturbance inputs or parameter variations are typical for the utilization of partial controllers of the order n_o-1 (Figs. 13, 14, 15). These have, however, been kept small by designing the controller in accordance with the requirement (2.138). The offset is scarcely noticeable above all in the case of the disturbance behavior (Figs.14, 15).

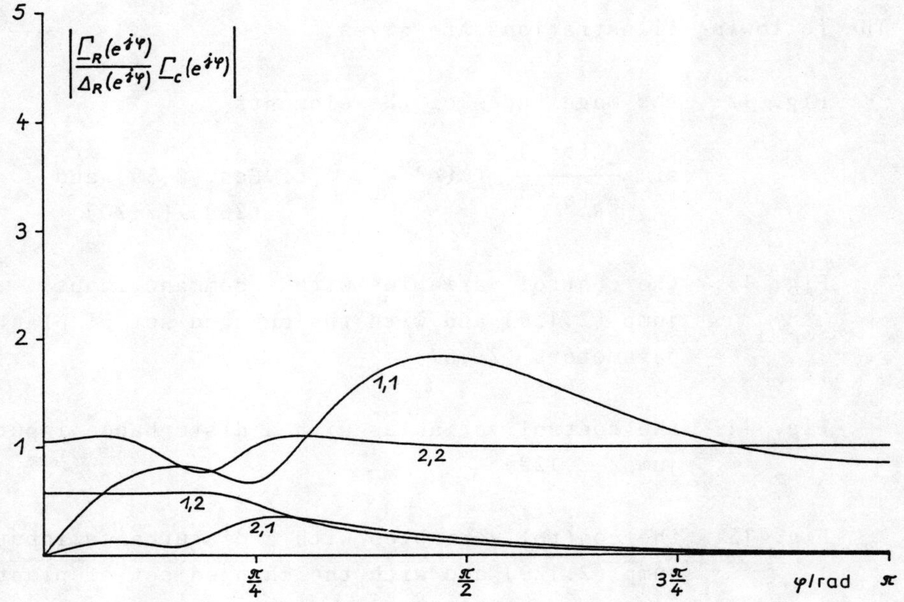

Fig.2.12: Magnitude of $\underline{\Gamma}_R\underline{\Gamma}_C/\Delta_R$; m = 3

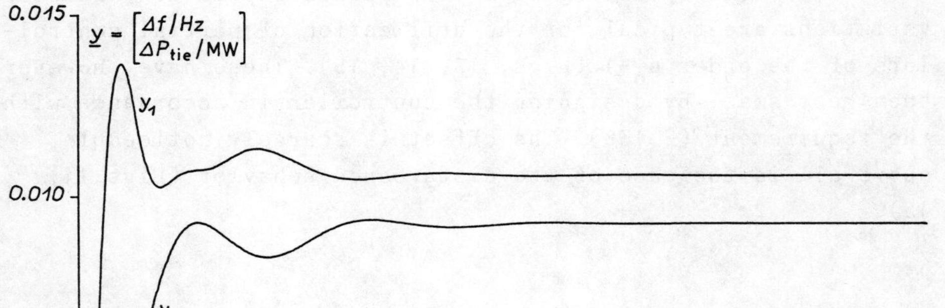

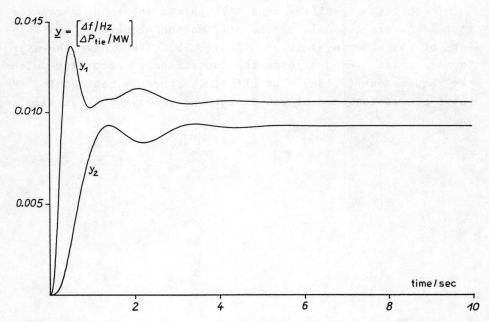

Fig.2.13: Command behavior with the changed plant parameter set (2.137); m = 3

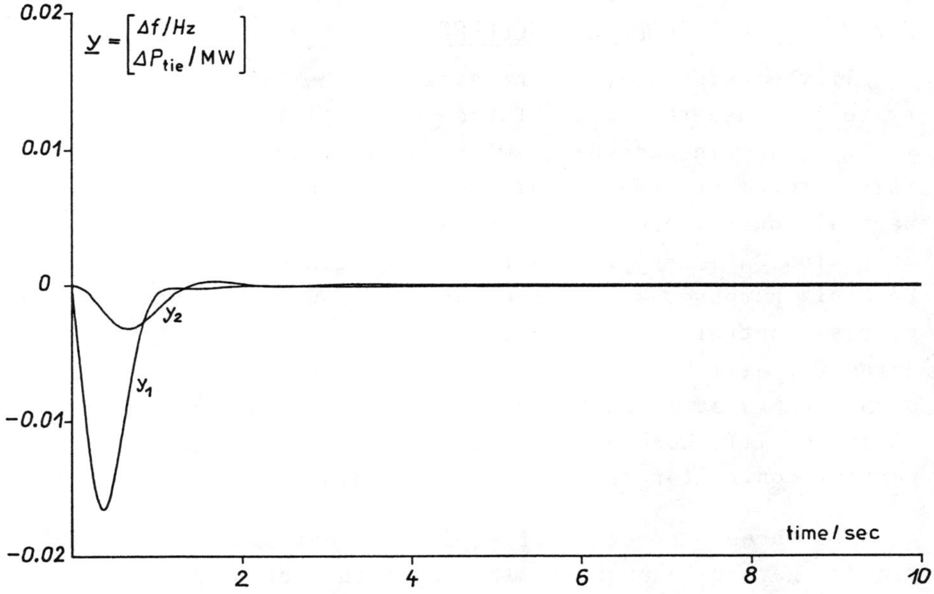

Fig.2.14: Disturbance behavior (nominal case); m = 3

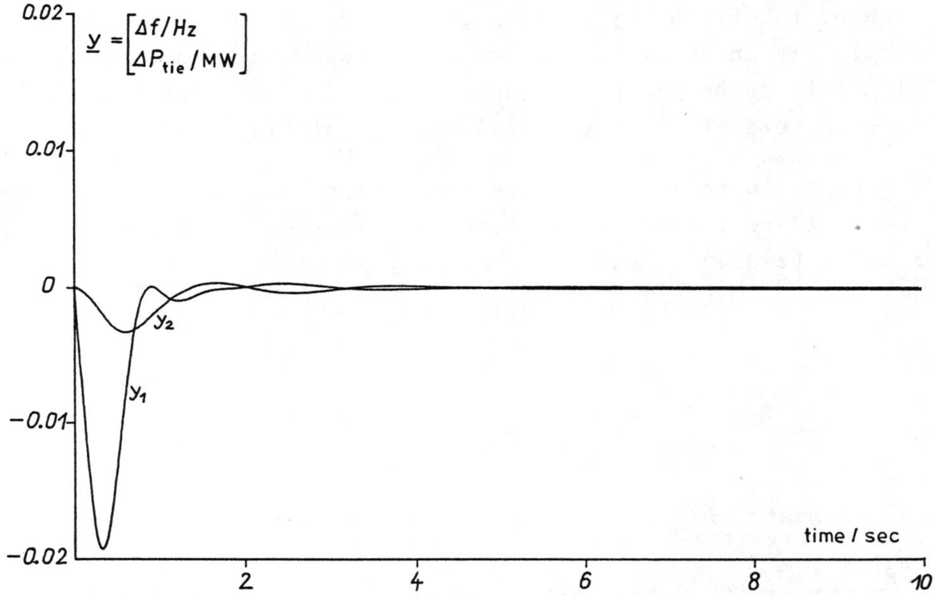

Fig.2.15: Disturbance behavior with the changed plant parameter set (2.137); m = 3

2.7.2 2 x 4th ORDER CONTROLLERS

If the offsets are to be avoided, then it will be necessary to increase the order of the two partial controllers to $m = n_o = 4$. In accordance with Eq.(2.38), we shall thus obtain a total of three freely selectable parameters for each partial controller. These can be used for satisfying the condition $\underline{\Gamma}_C(1) = \underline{0}$ in order thus to eliminate the offset. For this purpose, p = 2 free parameters will be necessary per partial controller, cf. section 2.5.1 . The remaining free parameter will be used for minimizing $||\underline{\Gamma}_C||$ within a frequency range still to be determined. The effect of the minimization will, however, be small since only one parameter per partial controller is available for this purpose.

All three parameters of a partial controller are determined together through resolution of the set of linear equations (2.124), this being, in the present case, a system of 5th order (quantity of free parameters + quantity of rows of $\underline{L}_o^*$) in accordance with Eq.(2.115). The quantity and position of the frequency points (2.116) must also be indicated previously in order to determine the frequency range in which $||\underline{\Gamma}_C||$ is to be minimized (henceforth, the additional weighting factors $\alpha(\varphi_\nu)$ in (2.117a) will be placed equal to one).

Figs. 16 to 19 illustrate the results obtained with a four-fold real controller eigenvalue at $\lambda(\underline{F}) = 0.4$, i.e. $\Delta_B(z) = (z-0.4)^4$, and the frequency points

$$\begin{aligned} \varphi_1 &= 0.15 \\ \varphi_2 &= 0.30 \\ &\vdots \\ \varphi_{10} &= 1.50 \quad . \end{aligned} \tag{2.139}$$

The elements of $\underline{\Gamma}_C$, and thus also the elements of

$\dfrac{\underline{\Gamma}_R(e^{j\varphi})}{\Delta_R(e^{j\varphi})}\ \underline{\Gamma}_C(e^{j\varphi})$ will now begin in the zero point for $\varphi = 0$

(Fig.16) and no offsets will arise as a result of disturbance inputs and parameter variations (Figs. 17, 18, 19).

The maximum amplitudes during the transient response can be reduced by relocating the minimization range to the lower frequency range:

$$\begin{aligned} \varphi_1 &= 0.05 \\ \varphi_2 &= 0.10 \\ &\vdots \\ \varphi_{10} &= 0.50 \quad . \end{aligned} \tag{2.140}$$

$\underline{r}_R\underline{r}_C/\Delta_R$ will now (Fig.20) have somewhat smaller values in the lower frequency range (up to approx. $\varphi = 0.5$) but will then ascend more powerfully than in Fig.16. The maximum amplitudes in Figs. 21, 22, 23 are somewhat lower, but the transient response is rather less steady, than in Figs. 17, 18, 19.

In respect of the "clock error" (the integral above the frequency deviation Δf should be as small as possible), the disturbance behavior according to Fig. 22/23 ought to be more interesting than the behavior according to Fig. 18/19.

The requirements made (2.127) and (2.128) will in both cases be satisfied, even in the case of the changed set of plant parameters (2.137).

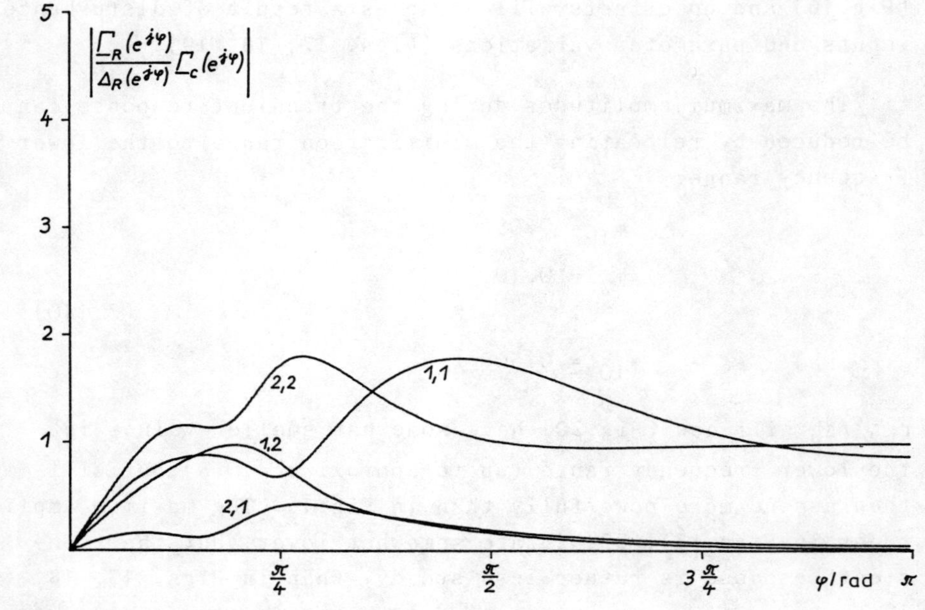

Fig.2.16: Magnitude of $\underline{\Gamma}_R\underline{\Gamma}_C/\Delta_R$; m = 4 .
Range of minimization: φ_{min} = (0, 1.5]

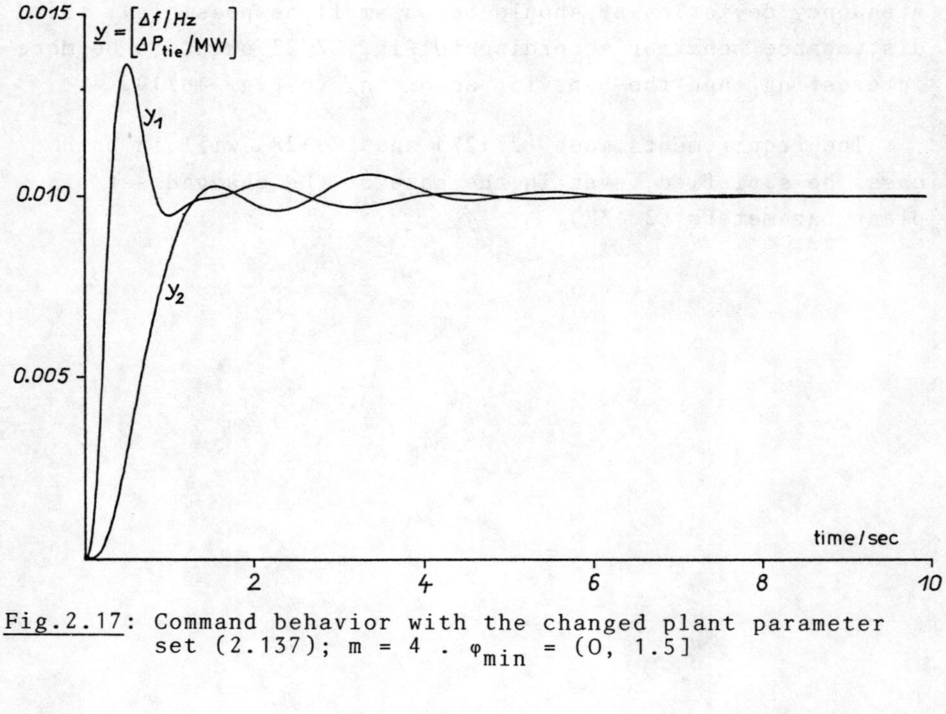

Fig.2.17: Command behavior with the changed plant parameter set (2.137); m = 4 . φ_{min} = (0, 1.5]

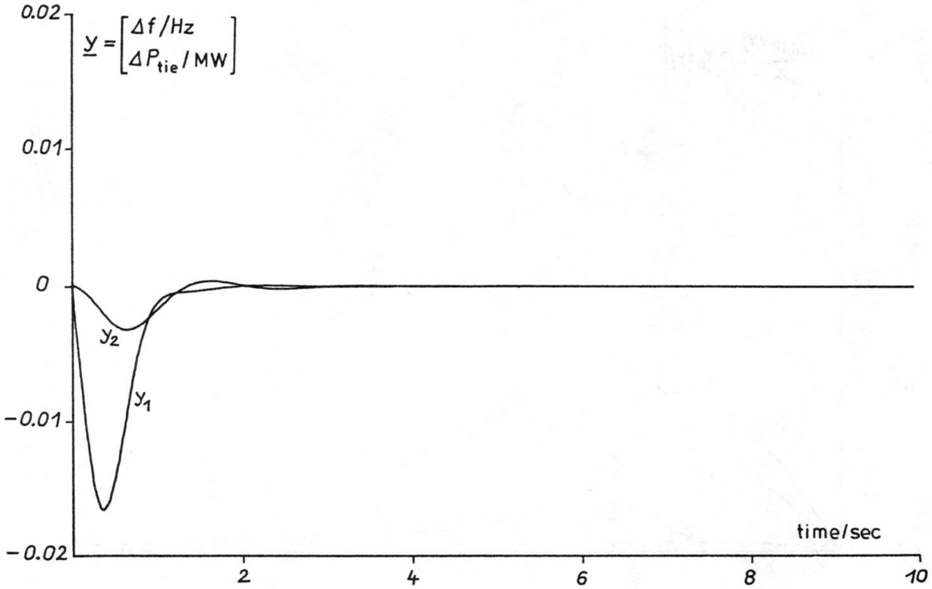

Fig.2.18: Disturbance behavior (nominal case); m = 4. φ_{min} = (0, 1.5]

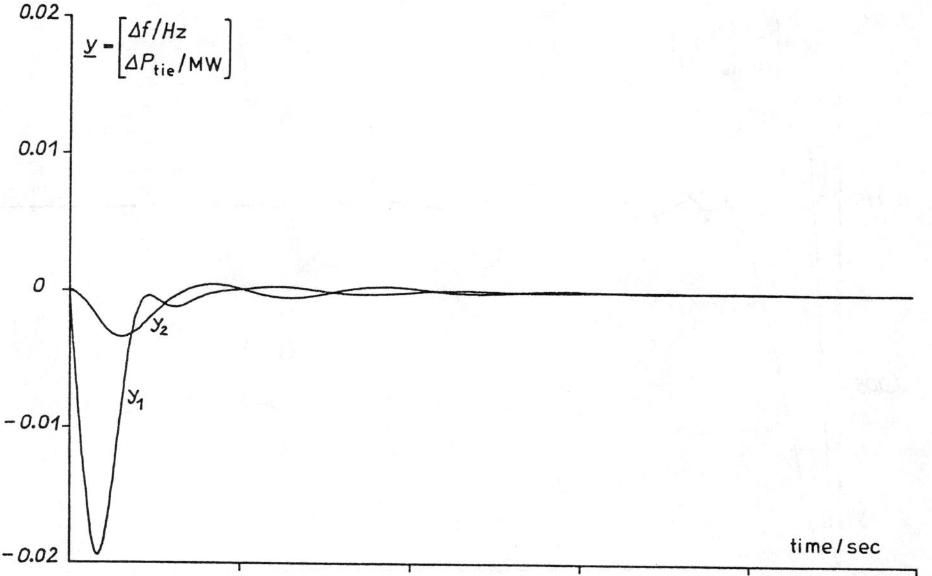

Fig.2.19: Disturbance behavior with the changed plant parameter set (2.137); m = 4. φ_{min} = (0, 1.5]

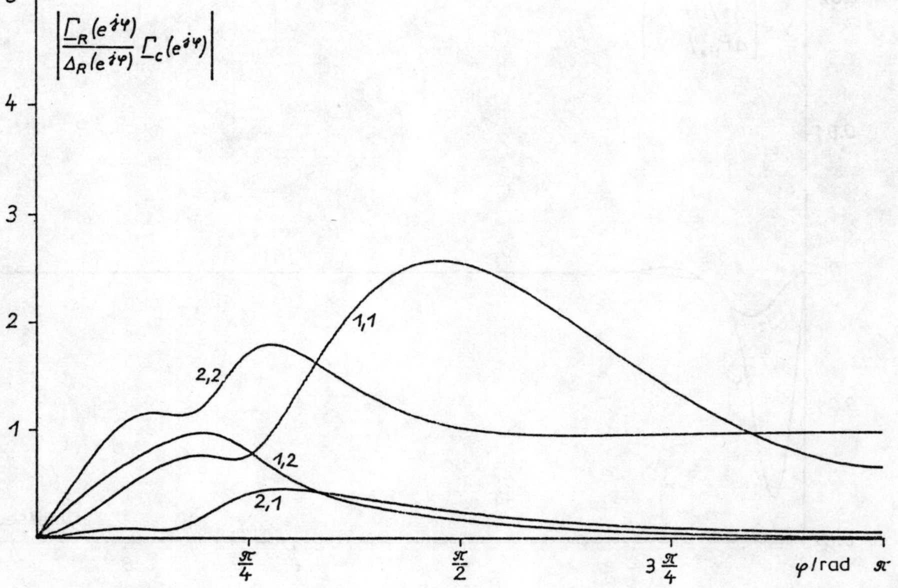

Fig.2.20: Magnitude of $\underline{\Gamma}_R\underline{\Gamma}_C/\Delta_R$; m = 4.
Range of minimization: φ_{min} = (0, 0.5]

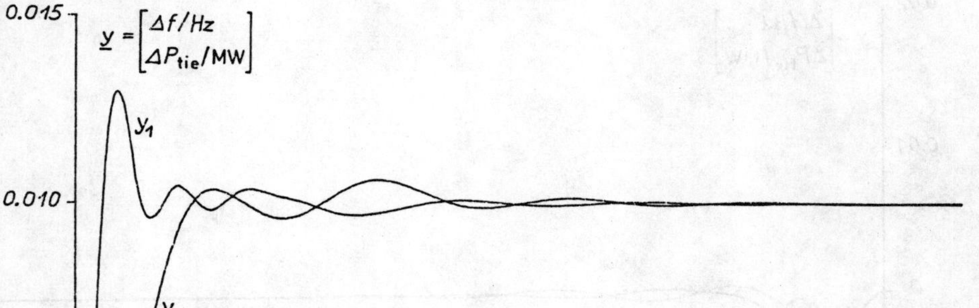

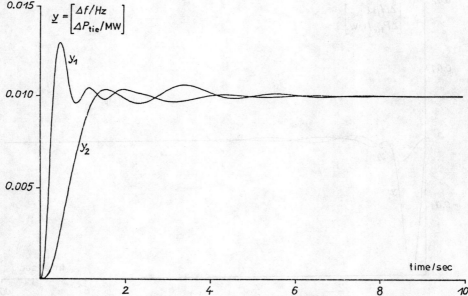

Fig.2.21: Command behavior with the changed plant parameter set (2.137); m = 4. φ_{min} = (0, 0.5]

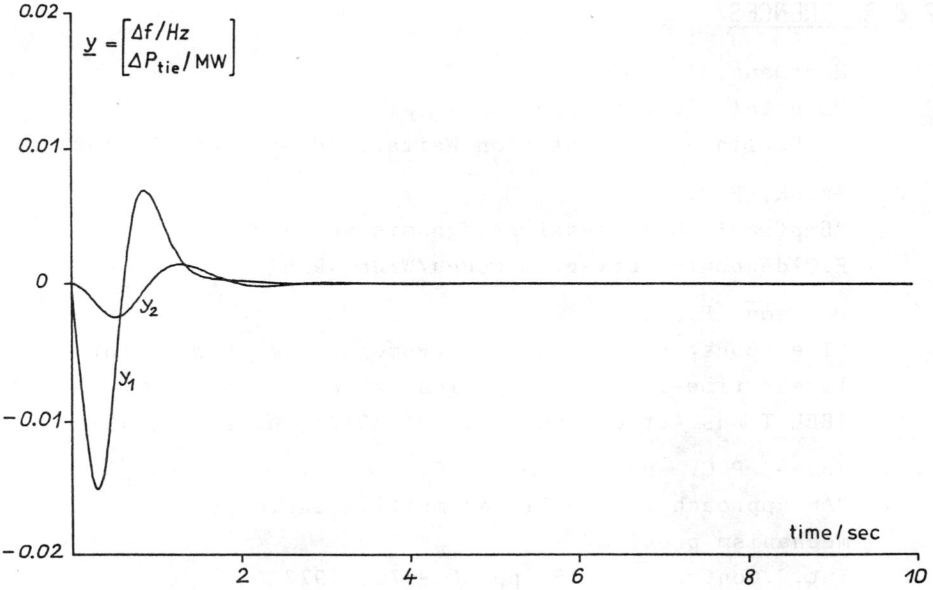

Fig.2.22: Disturbance behavior (nominal case); m = 4 φ_{min} = (0, 0.5]

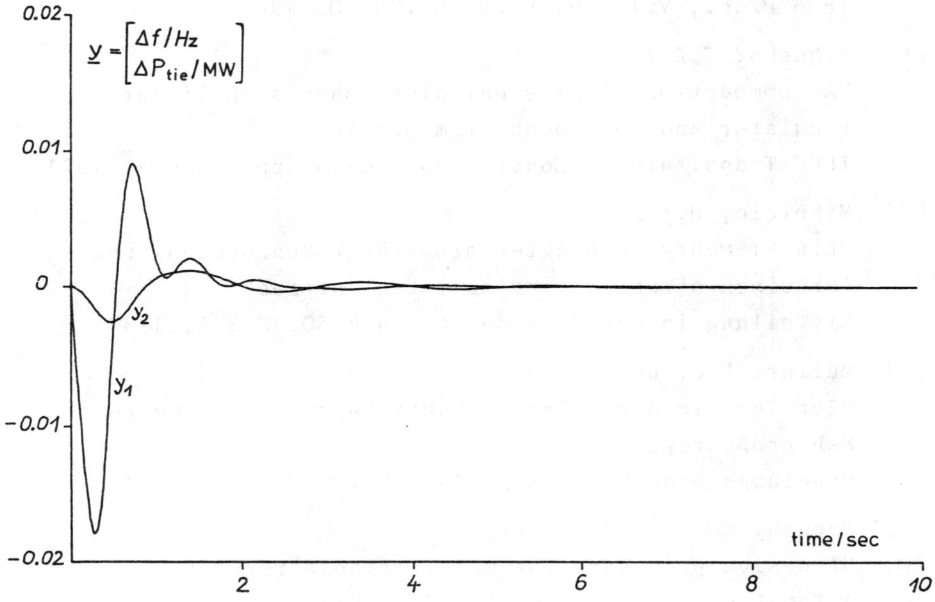

Fig.2.23: Disturbance behavior with the changed plant parameter set (2.137); m = 4. φ_{min} = (0, 0.5]

2.8 REFERENCES

[1] Hartmann, I.:
"Digitale Regelkreise"
TU Berlin - Dokumentation Weiterbildung, Heft 7, 1982

[2] Frank, P.M.:
"Empfindlichkeitsanalyse dynamischer Systeme"
R.Oldenbourg Verlag, München/Wien 1976

[3] Davison, E.J.:
"The robust control of a servomechanism problem for linear time-invariant multivariable systems"
IEEE Trans.Automat.Contr., Vol.AC-21, pp.25-34, 1976

[4] Young, P.C. and Willems, J.C.:
"An approach to the linear multivariable servo-mechanism problem"
Int.J.Contr., Vol.15, pp.961-979, 1972

[5] Noldus, E.:
"Disturbance rejection using dynamic output feedback"
IEEE Proc., Vol.129, Pt.D, pp.76-80, 1982

[6] Johnson, C.D.:
"Accommodation of external disturbances in linear regulator and servomechanism problems"
IEEE Trans.Automat.Contr., Vol.AC-16, pp.635-644, 1971

[7] Mikolcic, H.:
"Ein PI-Mehrgrößenregler mit Störgrößenaufschaltung für einen stromrichtergespeisten Antrieb"
Mitteilung in der Regelungstechnik 30, S.326, 1982

[8] Müller, P.C. und Lückel, J.:
"Zur Theorie der Störgrößenaufschaltung in linearen Mehrgrößenregelkreisen"
Regelungstechnik 25, S.54-59, 1977

[9] Sebakhy, O.A. and Wonham, W.M.:
"A design procedure for multivariable regulators"
Automatica, Vol.12, pp.467-478, 1976

[10] Luenberger, D.G.:
"Observers for multivariable systems"
IEEE Trans.Automat.Contr., Vol.AC-11, pp.190-197, 1966

[11] Fortmann, T.E. and Williamson, D.:
"Design of low-order observers for linear feedback control laws"
IEEE Trans.Automat.Contr., Vol.AC-17, pp.301-308, 1972

[12] Luenberger, D.G.:
"Canonical forms for linear multivariable systems"
IEEE Trans.Automat.Contr., Vol. AC-12, pp.290-293, 1967

[13] Hartmann, I.:
"Algorithmen in dynamischen Systemen"
TU Berlin, Vorlesungsskript, 1976

[14] Fosha, C.E. and Elgerd, O.I.:
"The megawatt-frequency control problem: a new approach via optimal control theory"
IEEE Trans.Power App.Sys., Vol.PAS-89, pp.563-571, 1970

[15] Poltmann, R.:
"Ein optimaler Empfindlichkeitsentwurf für lineare zeitdiskrete Mehrgrößenregelkreise"
Dissertation TU Berlin, 1984

3 Robust Design of Multimodel Feedback Systems

3.1 POLE-DISTANCE-DESIGN

3.1.1 INTRODUCTION

In this chapter a design procedure for multimodel multivariable control systems is described where one constant controller for all members of the multimodel-family is designed. The objective of this procedure is to adapt the closed-loop behaviour of the different models of the multimodel-family to an required behaviour. Here the required behaviour is described by a pole-configuration of the closed-loop system. One can get this desired pole-configuration for instance by a single controller-design for each variation-case of the multi-model-family. If this controller fulfils the given static and dynamic requirements, then the belonging closed-loop poles are used as desired pole-configuration. Normally the controllers in the different variation cases are different. The design-procedure described in the sequel determines one constant controller for all variation cases, so that the desired closed-loop pole configurations are attained as good as possible. Therefore a pole-distance is introduced. Using this pole-distance it is possible to describe the controller-design as an extrem-value-search. In the design the different inputs of the plant may be determined by using the controllability coefficients that give us a hint of small input variables, further information see Lange [6].

3.1.2 DEFINITION OF THE POLE-DISTANCE

Given there are two pole-configurations a and b with n poles in each configuration. The vector v contains the poles of a and the vector $\underline{w}$ contains the poles of b. The arrangement of the poles in $\underline{w}$ runs through all permutations p, p = 1...n!.

A permutated arrangement of the elements of $\underline{w}$ is denoted $\underline{w}_p$. The pole distance between the pole-configuration a and b is

$$d = \min_p [\underline{w}_p - \underline{v}]^* [\underline{w}_p - \underline{v}]$$

(* conjugate, transposed; p ist the p-th permutation of the arrangement in $\underline{w}$).

Note that the pole-distance is not the distance between two poles. The following example illustrates the determination of the pole distance. The pole configuration a consists of the poles $v_1 = -6$, $v_2 = j2$, $v_3 = -j2$ and the configuration b consists of the poles $w_1 = -8$, $w_2 = -2+j$, $w_3 = -2-j$. Now all possible assignments between the poles of configuration a and b are made as shown in figures 3.1. For each assignment the sum of the squares of the single distances are determined. The minimum out of this sums is equal to the pole distance.

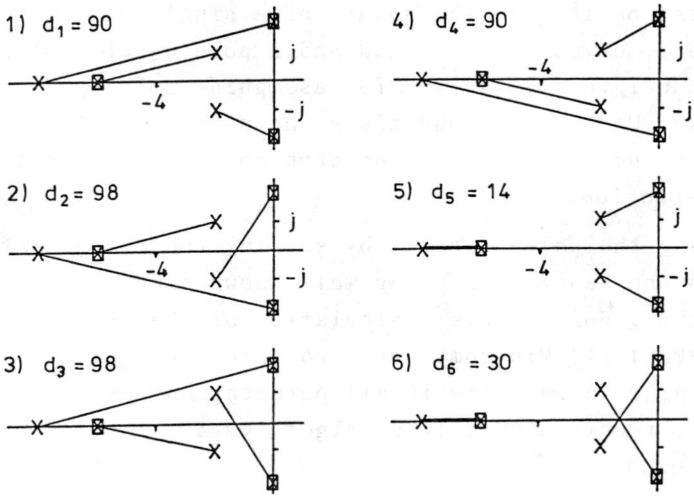

Fig.3.1: The 6 permutations of the example. The pole-distance is given by permutation 5.

3.1.3 POLE-DISTANCE AS ASSIGNMENT-PROBLEM

Assignment-problem: Suppose there is a transport-society with n garages at n different places. There is one car in each garage. These n cars have to go to n different destinations, one car to each destination. Now a driving-schedule for minimizing the costs is required. Therefore a cost-matrix is built up. The row-index of this matrix is the index of the place of the garage. The column index is the index of the destination. The value of the matrixelement with index i, j is equal to the expenses if one car drives from garage i to destination j. If this expense is proportional to the distance between garage and destination, the costmatrix contains only the distances between the garages and the destinations. The problem is now to find n elements of the costmatrix whose sum is minimal and no two elements belong to the same row or the same column. Such problems are called assignment-problems.

The poles of configuration a are now the starting-points ("garages") and the poles of configuration b are now the end-points ("destinations"). The square of a single distance between a pole v_i of configuration a and a pole w_j of configuration b is equal to the "costs" for assigning pole v_i to pole w_j (fig.3.2). Under this conditions the problem of finding the pole-distance is equal to the problem of solving the assignment-problem.

Determing the pole-distance by solving the assignment-problem has the benefit of using well-known time saving computer-algorithms. For example: Calculation of the pole-distance with a DEC-PDP11/44 Minicomputer, two pole configurations with 9 poles each, required time if all permutations are determined: 530 seconds; required time if an algorithm is used from [1]: 80 milliseconds.

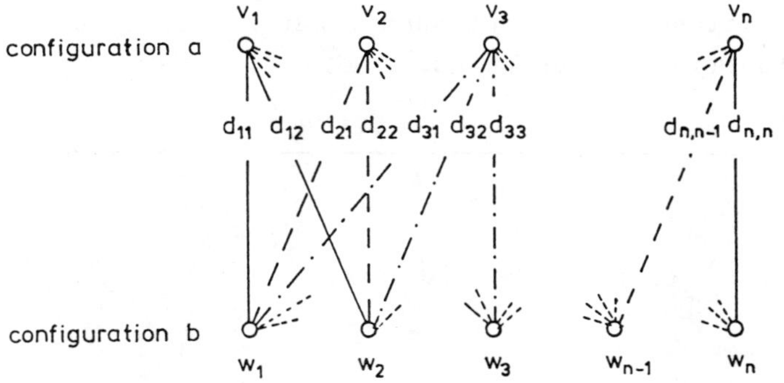

Fig.3.2: The pole-distance determination as assignment-problem.

3.1.4 CONTINUITY OF THE POLE-DISTANCE

The distance for a permutation p is defined by

$$d_p = [\underline{w}_p - \underline{v}]^* [\underline{w}_p - \underline{v}]$$

(* conjugate-komplex, transposed; p is the p-th permutation). Each of this p = n! distances can be observed as a distance-function $d_p = f(\underline{w}_p)$.

All this distance-functions are continuous in the whole complex plane. The minimum out of all this distance-functions changes from function i to function j only, if these two functions cross each other. Both functions are continuous at the intersection. The transition of the minimum from one distance-function to another distance-function is therefore continuous (fig.3.3 and 3.4). The pole-distance is normally not continuous differentiable.

Summary of the pole-distance properties:

- if the two configurations in question have the same poles the pole-distance is zero
- if the configurations a relatively a' and b relatively b' are the same, but the poles in v relatively v' and w relatively w' are otherwise ordered, then the pole distance between a,b and a',b' is equal

- the pole distance is always greater or equal to zero
- the pole-distance is a continuous, but not continuous differentiable function of the poles.

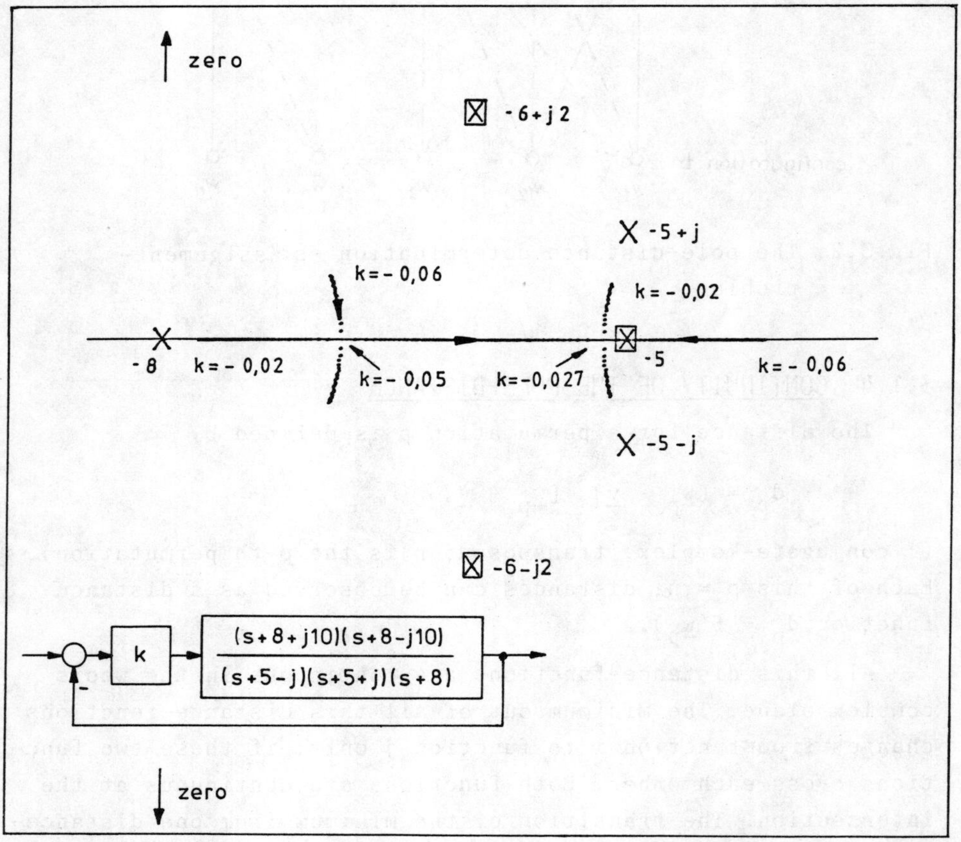

Fig.3.3: The root locus of the closed-loop circuit in the down left corner for k = -0.06 to k = -0.02. (x:plant-poles; ⊠ : to this poles the pole distance is determined).

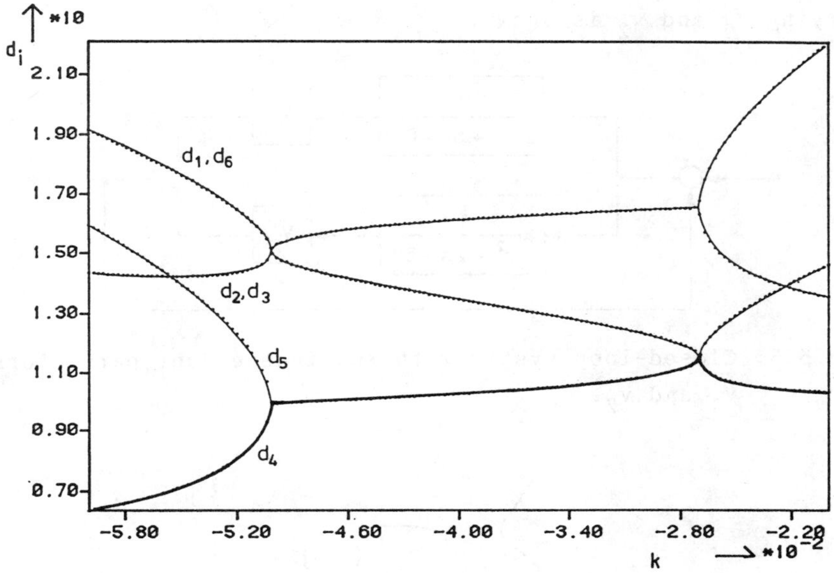

Fig.3.4: The 6 distance-functions of the closed-loop configuration in example fig. 3.3.

In figure 3.4 the bold line is the pole-distance in respect to varying k. The transition of the pole-distance from one distance-function to another distance-function is continuous.

3.1.5 THE POLE-DISTANCE AS A FUNCTION OF THE CONTROLLER-PARAMETERS

The following example shows what function the pole-distance can be, if some of the controller-parameters vary. Given is the feedback-system of fig.3.5 with varying v_1 and v_2. In fig.3.5 the root-locus for the closed loop-system is shown with $v_1 = 2$ and v_2 varying between 0 and 10. In fig. 3.7a detail of the root-locus for varying v_1 and varying v_2 is shown. The 0 in this figure denotes the poles, for which the distance to the varying closed loop-poles is determined. The □ in this figure denotes the closed-loop poles for which this distance is minimal.

In the next figure 3.8 the inverse of the pole distance for varying v_1 and v_2 is shown.

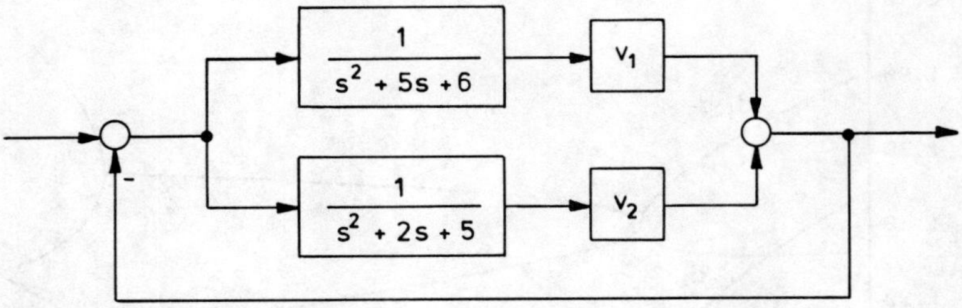

Fig.3.5: Closed-loop system with two independent parameters v_1 and v_2.

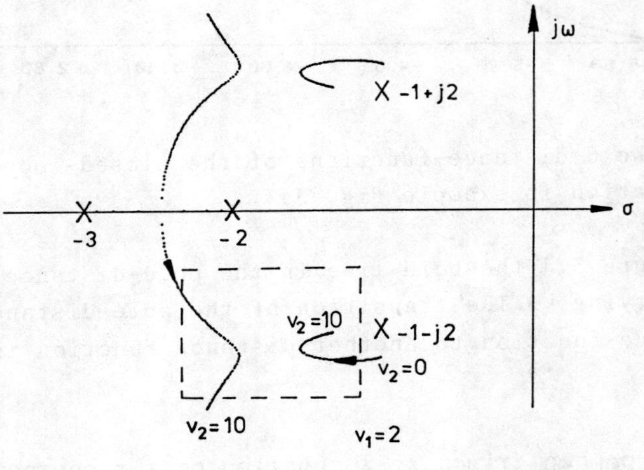

Fig.3.6: The root-locus of the control-system from fig.3.5 for $v_1 = 2$ and $v_2 = 0....10$. The area denoted by--- is shown in fig.3.7 for varying v_1 and v_2.

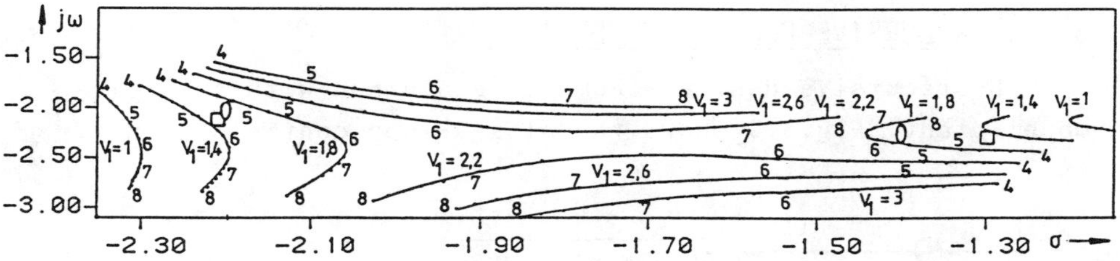

Fig.3.7: The root-locus of the control-system from fig.3.6 for $v_1 = 1...3$ and $v_2 = 4...8$.

The desired closed-loop poles are denoted by O in fig.3.7, they are lying at $s_{1,2} = -2,2 \pm j2$ and $s_{3,4} = -1,4 \pm j2,2$. The pole-distance minimum is $d = 0,067$ for $v_1 = 1.5$ and $v_2 = 5.2$. The closed-loop poles in the pole-distance minimum are $s_{5,6} = -1.29 \pm j2.31$ and $s_{7,8} = -2.21 \pm j2.1$.

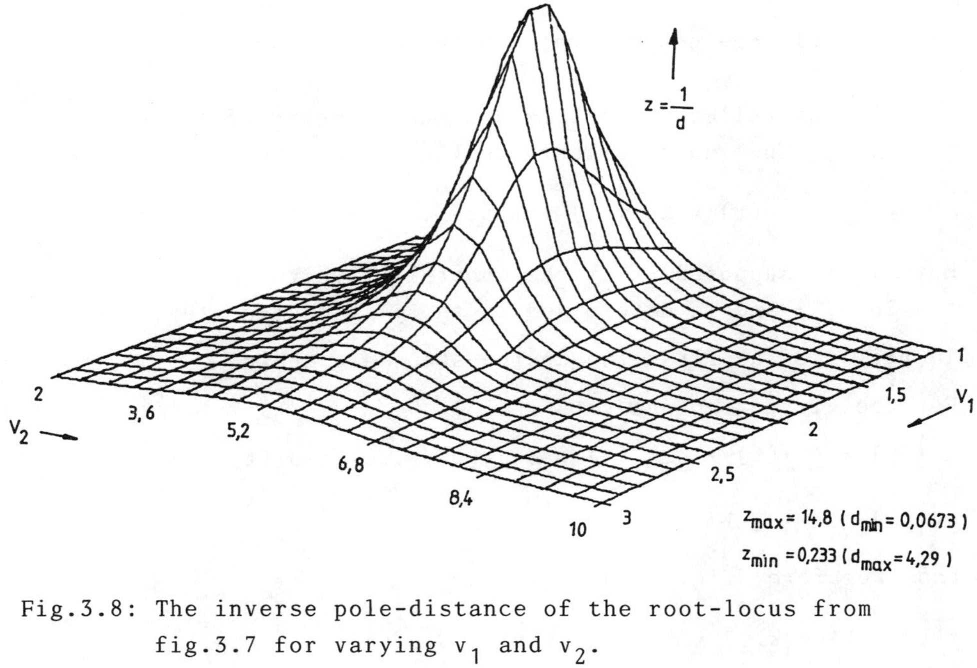

Fig.3.8: The inverse pole-distance of the root-locus from fig.3.7 for varying v_1 and v_2.

3.1.6 RECURSIVE CONTROLLER DESIGN

The recursive design scheme is explained with a single-model-plant. Fig.3.9 shows the closed-loop control circuit.

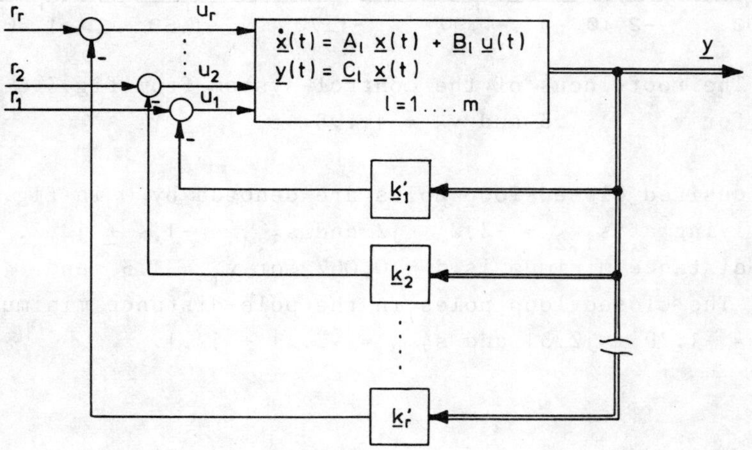

Fig.3.9: Closed-loop control-system.

The controller parameters are the elements of the vectors $\underline{k}_1,\dots,\underline{k}_r$. The input variable in the design-step j is

$$u_j(t) = \underline{k}_j' \, \underline{y}(t) + r_j(t) \ .$$

Here it is supposed, that the controller-vector $\underline{k}_j$ belongs to the input j and is determined in the j-th design-step.

In fig.3.10 the first two design-steps are shown.

For the first step one gets

$$\dot{\underline{x}}(t) = \underline{A}\ \underline{x}(t) + \underline{b}_1 u_1(t) + \underline{b}_2 u_2(t) + \dots + \underline{b}_r u_r(t)$$

$$u_1(t) = -\underline{k}_1' \, \underline{y}(t) + r_1(t)$$

and therefore, fig.3.9 and 3.10,

$$\dot{\underline{x}}(t) = [\underline{A} - \underline{b}_1 \underline{k}_1' \underline{C}]\ \underline{x}(t) + \underline{b}_1 r_1(t) + \underline{b}_2 u_2(t) + \dots + \underline{b}_r u_r(t) \ .$$

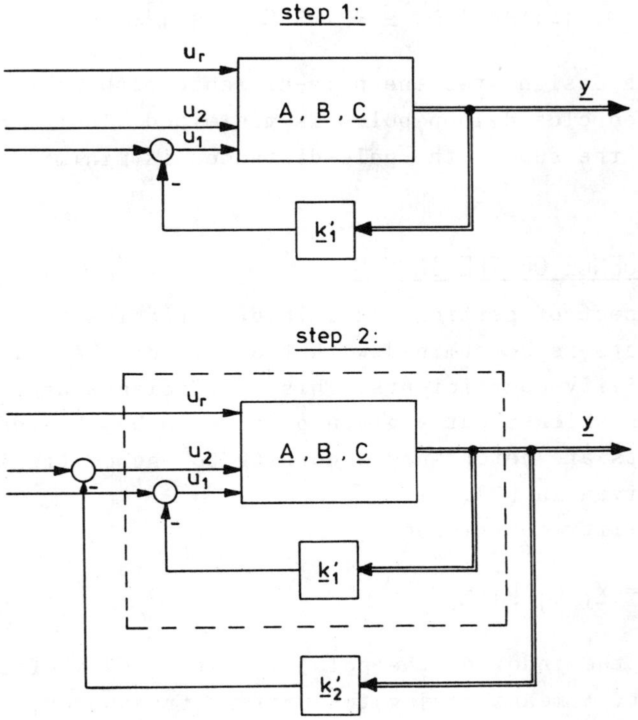

Fig.3.10: The first two design-steps.

In the second design-step one starts with the new plant

$$\dot{\underline{x}}(t) = \underline{\tilde{A}}_1 \underline{x}(t) + \underline{B}\, \underline{\tilde{u}}(t)$$

with the input vector

$$\underline{u}(t) = \begin{bmatrix} r_1(t) \\ u_2(t) \\ \vdots \\ u_r(t) \end{bmatrix}$$

The systemmatrix after the j-th design step is

$$\underline{\tilde{A}}_j = \underline{\tilde{A}}_{j-1} - \underline{b}_j\, \underline{k}_j'\, \underline{C}$$

$$\underline{\tilde{A}}_o = \underline{A}$$

and the characteristic polynomial after the j-th step is

$$\Delta_j(s) = \Delta_{j-1}(s) + \underline{k}_j' \, \underline{\underline{C}} \, \mathrm{adj}[s \, \underline{\underline{E}} - \tilde{A}_{j-1}] \underline{b}_j \, .$$

During each design step the pole-distance between the desired and attained closed-loop poles is minimized. In the multi-model-case the sum of the pole-distances is minimized.

3.1.7 SEQUENCE OF THE INPUTS

In respect of getting small input variables the sequence of the inputs is determined with the help of the following controllability-coefficients. This coefficients are used to decide with which input a given pole is to be shifted. This coefficients are determined according to the controllability measures given in [2].
Controllability coefficient:

$$s_{ij} = \underline{x}_i^* \, \underline{b}_j \, \underline{b}_j' \, \underline{x}_i$$

i is the index of the pole, j is the index of the input, * means conjugate-complex, transposed; x_i is the left-eigenvector i of the plant-systemmatrix, b_j is the j-th column of the input-matrix.

At the beginning of each recursion-step a table of all controllability-coefficients is determined. For a given pole, which has to be shifted, those input is used for which the according controllability-coefficient has the largest value.

3.1.8 MINIMIZING THE POLE-DISTANCE

Normally the pole-distance is not continuous differentiable and therefore one cannot use gradient-methods for finding the minimal pole-distance. Hence the minimum of the pole-distance is determined with help of the Rosenbrook method. The principal way of this method is as follows:

From a starting point a one dimensional extrem value-search in the n orthogonal directions are begun. The search step-width and search-direction is controlled by the success or failure of the search. The one-directional search is continued until there is no decreasing of the pole-distance neither in forward nor in backward-direction. Then the search is continued in the next orthogonal direction, starting from the last best-point and so on until all of the orthogonal directions are passed through. Now a new set of orthogonal directions is determined and the first in all lies on the connection between the starting point of the first search and the last best-point.

3.1.9 EXAMPLE

The plant is a light-helicopter with allweather capability see fig.3.11. For a complete model description compare [3]. There are 4 different linear, timeinvariant models of order 8. This models are found for forward-speed of 40, 60, 80 and 100 knots. The state variables are the translatoric speeds u, v, w in x, y, z directions, the three angular-velocities, the pitch-and the roll-angle. The plant inputs are the longitudinal cyclic control, the collective pitch control, the lateral cyclic control and the pedal. The three translatoric speeds are not measured and the other state variables could be used for the feedback.

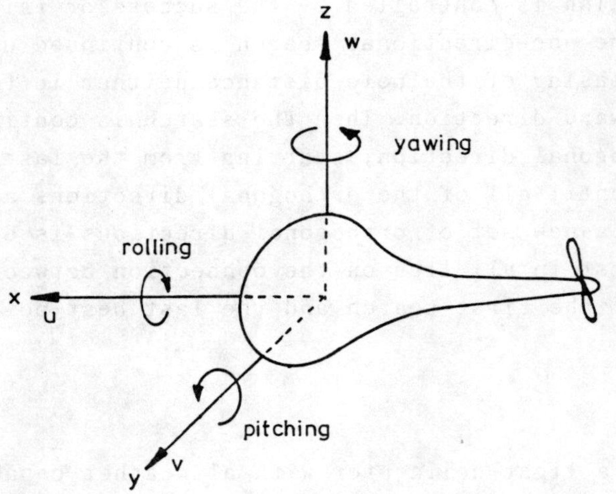

Fig.3.11: Helicopter model [3].

We obtain the linear timeinvariant state-equation

$$\dot{\underline{x}}(t) = \underline{A}\ \underline{x}(t) + \underline{B}\ \underline{u}(t)$$

with the eight states

$$\underline{x}^T = [\Delta u,\ \Delta w,\ q,\ \Delta\theta,\ v,\ p,\ \Delta\Phi, r]$$

u,v,w : translatoric speed

p,q,r : angular velocity

θ : pitch-angle
Φ : roll-angle } Euler-angles ,

here Δ denotes the deviation from trim-values, and the four inputs of the plant-model are called

$u_1 := \delta_e$: longitudinal cyclic-control

$u_2 := \delta_c$: collective pitch control

$u_3 := \delta_a$: lateral cyclic control

$u_4 := \delta_p$: pedal

The data of the four models (differentforward speeds) are contained in table 3.1.

$\underline{A}_{40}$ =

	ΔU	ΔW	Q	Δθ	ΔV	P	ΔΦ	R
	-0.24E-01	0.17E-01	0.58E+00	-0.32E+02	0.12E-02	0.12E+01	0.00E+00	-0.10E+00
	-0.89E-01	-0.56E+00	0.69E+02	-0.85E+00	0.11E-02	0.31E+01	0.10E+01	-0.84E+03
	0.14E-01	-0.17E-02	-0.37E+01	0.00E+00	-0.25E-02	-0.27E+00	0.00E+00	0.12E+00
	0.00E+00	0.00E+00	0.99E+00	0.00E+00	0.00E+00	0.00E+00	0.00E+00	0.32E-01
	0.38E-02	-0.13E-02	0.11E+01	0.27E-01	-0.10E+00	-0.92E+00	0.32E+02	-0.66E+02
	-0.41E-02	-0.22E-03	-0.17E-01	0.00E+00	0.62E-01	-0.96E+01	0.00E+00	0.16E-01
	0.00E+00	0.00E+00	-0.86E-03	0.00E+00	0.00E+00	0.10E+01	0.00E+00	0.26E-01
	-0.61E-02	-0.66E-02	0.16E+00	0.00E+00	0.30E-01	0.78E-01	0.00E+00	-0.65E+00

$\underline{B}_{40}$ =

	δ_e	δ_c	δ_a	δ_p
	-0.76E+00	0.21E+00	0.37E-03	0.20E-01
	-0.15E+01	-0.89E+01	0.25E+00	0.29E-01
	0.10E+01	0.15E+00	0.63E-01	0.20E-01
	0.00E+00	0.00E+00	0.00E+00	0.00E+00
	-0.50E-03	-0.11E+00	0.72E+00	-0.16E+01
	-0.19E+00	-0.12E+00	0.23E+01	-0.97E+00
	0.00E+00	0.00E+00	0.00E+00	0.00E+00

$\underline{A}_{60}$ =

-.32E-01	.19E-01	.18E 01	-.32E 02	.17E-02	.12E 01	.00E 00	-.54E-01
-.43E-01	-.73E 00	.10E 03	-.20E 00	-.34E-02	.4oE 01	.96E 00	-.84E-01
.13E-01	.53E-03	-.38E 01	.00E 00	-.24E-02	.31E 00	.00E 00	.11E 00
.00E 00	.00E 00	.99E 00	.00E 00	.00E 00	.00E 00	.00E 00	.30E-01
.12E-02	-.36E-02	.12E 01	.60E-02	-.14E 00	-.21E 01	.32E 02	-.10E 03
-.54E-02	-.10E-02	.13E-01	.00E 00	-.68E-01	-.96E-01	.00E 00	.15E 00
.00E 00	.00E 00	.18E-03	.00E 00	.00E 00	.10E 01	.00E 00	.62E-02
-.38E-02	-.96E-02	.24E 00	.00E 00	.39E-01	.15E 00	.00E 00	-.85E 00

$\underline{B}_{60}$ =

-.74E 00	.12E 00	-.13E-01	.27E-01
-.22E 01	-.93E 01	.40E 00	.39E-01
.10E 01	.28E 00	.62E-01	.21E-01
.00E 00	.00E 00	.00E 00	.00E 00
-.55E-02	-.20E-01	.71E 00	-.17E 01
-.18E 00	-.83E-01	.23E 01	-.10E 01
.00E 00	.00E 00	.00E 00	.00E 00
-.36E-01	.10E 00	.29E-01	.15E 01

Table 3.1: Data for helicopter-model.

$$\underline{A}_{80} = \begin{bmatrix} \text{-.40E-01} & \text{.18E-01} & \text{.52E 01} & \text{-.32E 02} & \text{.21E-02} & \text{.12E 01} & \text{.00E 00} & \text{.50E-01} \\ \text{-.16E-01} & \text{-.83E 00} & \text{.13E 03} & \text{.66E 00} & \text{-.71E-02} & \text{.51E-01} & \text{.10E 01} & \text{-.10E 00} \\ \text{.13E-01} & \text{.34E-02} & \text{-.39E 01} & \text{.00E 00} & \text{-.24E-02} & \text{-.35E 00} & \text{.00E 00} & \text{.11E 00} \\ \text{.00E 00} & \text{.00E 00} & \text{.99E 00} & \text{.00E 00} & \text{.00E 00} & \text{.00E 00} & \text{.00E 00} & \text{.32E-01} \\ \text{.37E-03} & \text{-.45E-02} & \text{.12E 01} & \text{-.21E-01} & \text{-.17E 00} & \text{-.55E 01} & \text{.32E 02} & \text{-.13E 03} \\ \text{-.59E-02} & \text{-.18E-02} & \text{.26E-01} & \text{.00E 00} & \text{-.75E-01} & \text{-.95E 01} & \text{.00E 00} & \text{.27E 00} \\ \text{.00E 00} & \text{.00E 00} & \text{.66E-03} & \text{.00E 00} & \text{.00E 00} & \text{.10E 01} & \text{.00E 00} & \text{-.20E-01} \\ \text{-.30E-02} & \text{-.98E-02} & \text{.29E 00} & \text{.00E 00} & \text{.46E-01} & \text{.19E 00} & \text{.00E 00} & \text{-.10E 01} \end{bmatrix}$$

$$\underline{B}_{80} = \begin{bmatrix} \text{-.73E 00} & \text{.49E-01} & \text{-.11E-01} & \text{.41E-01} \\ \text{-.30E 01} & \text{-.99E 01} & \text{.57E 00} & \text{.53E-01} \\ \text{.10E 01} & \text{.41E 00} & \text{.36E 00} & \text{.25E-01} \\ \text{.00E 00} & \text{.00E 00} & \text{.00E 00} & \text{.00E 00} \\ \text{-.84E-02} & \text{-.51E-01} & \text{.72E 00} & \text{-.19E 01} \\ \text{-.18E 00} & \text{-.11E 00} & \text{.23E 01} & \text{-.11E 01} \\ \text{.00E 00} & \text{.00E 00} & \text{.00E 00} & \text{.00E 00} \\ \text{-.40E-01} & \text{.20E 00} & \text{.26E-01} & \text{.16E 01} \end{bmatrix}$$

$$\underline{A}_{100} = \begin{bmatrix} \text{-0.49E-01} & \text{0.19E-01} & \text{0.11E+02} & \text{0.32E+02} & \text{0.26E-02} & \text{0.11E+01} & \text{0.00E+00} & \text{0.28E+00} \\ \text{0.13E-03} & \text{-0.90E+00} & \text{0.17E+03} & \text{0.17E+01} & \text{-0.11E-01} & \text{0.62E+01} & \text{0.12E+01} & \text{-0.14E+00} \\ \text{0.14E-01} & \text{0.61E-02} & \text{-0.40E+01} & \text{0.00E+00} & \text{-0.22E-02} & \text{-0.40E+00} & \text{0.00E+00} & \text{0.11E+00} \\ \text{0.00E+00} & \text{0.00E+00} & \text{0.99E+00} & \text{0.00E+00} & \text{0.00E+00} & \text{0.00E+00} & \text{0.00E+00} & \text{0.37E-01} \\ \text{0.37E-03} & \text{-0.63E-02} & \text{0.12E+01} & \text{-0.64E-01} & \text{-0.21E+00} & \text{0.11E+02} & \text{0.32E+02} & \text{-0.16E+03} \\ \text{-0.60E-02} & \text{-0.34E-02} & \text{0.29E-01} & \text{0.00E+00} & \text{-0.82E-01} & \text{-0.93E+01} & \text{0.00E+00} & \text{0.33E+00} \\ \text{0.00E+00} & \text{0.00E+00} & \text{0.20E-02} & \text{0.00E+00} & \text{0.00E+00} & \text{0.10E+01} & \text{0.00E+00} & \text{0.53E-01} \\ \text{-0.29E-02} & \text{-0.59E-02} & \text{0.29E+00} & \text{0.00E+00} & \text{0.50E-01} & \text{0.17E+00} & \text{0.00E+00} & \text{-0.11E+01} \end{bmatrix}$$

$$\underline{B}_{100} = \begin{bmatrix} \text{-0.73E+00} & \text{-0.14E-01} & \text{0.27E-02} & \text{0.67E-01} \\ \text{-0.39E+01} & \text{-0.10E+02} & \text{0.74E+00} & \text{0.72E-01} \\ \text{0.10E+01} & \text{0.55E+00} & \text{0.47E-01} & \text{0.35E-01} \\ \text{0.00E+00} & \text{0.00E+00} & \text{0.00E+00} & \text{0.00E+00} \\ \text{0.14E-02} & \text{-0.69E-01} & \text{0.73E+00} & \text{-0.20E+01} \\ \text{-0.17E+00} & \text{-0.14E+00} & \text{0.23E+01} & \text{-0.12E+01} \\ \text{0.00E+00} & \text{0.00E+00} & \text{0.00E+00} & \text{0.00E+00} \\ \text{-0.28E-01} & \text{0.25E+00} & \text{0.21E-01} & \text{0.17E+00} \end{bmatrix}$$

Design objectives

The closed loop poles for all four cases have to be shifted into "good"-or "acceptable"-regions. These regions are defined by pilots. The "slow" (fig.3.12b) and "fast" (fig.3.12a) conjugate-complex pole-pairs have their own different regions. In figure 3.12 the defined pole regions are plotted as dotted areas. The desired closed loop poles, see table 3.2, are different for the four cases, because it is difficult to achieve the same closed-loop pole-configuration for all four cases. In figure 3.12a there are two plant-poles on the real axis at -4 and -0.8, which coincide with desired poles.

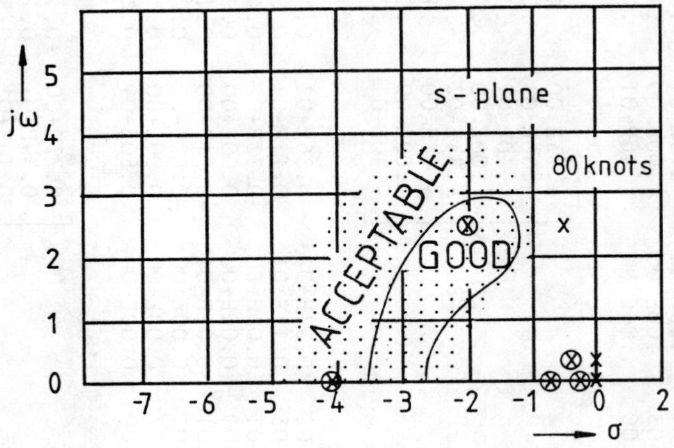

Fig.3.12a: Plant-[x] and desired poles [⊗] , here "fast" pole-area are dotted.

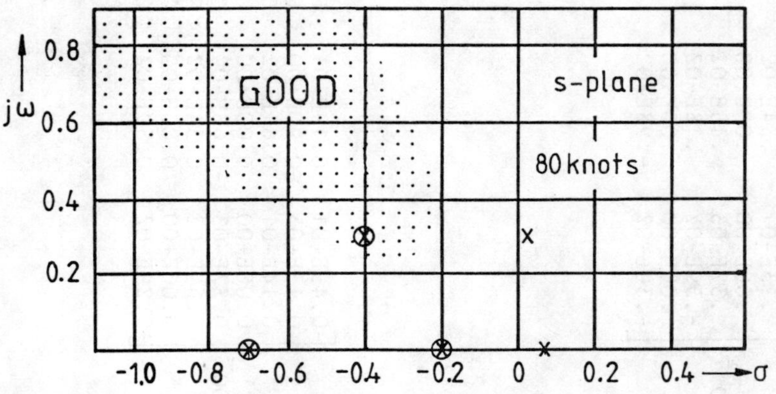

Fig.3.12b: Plant-[x] and desired poles [⊗] , here "slow" pole area are dotted.

	40 knots		60 knots		80 knots		100 knots	
	Real	Imag	Real	Imag	Real	Imag	Real	Imag
Open loop poles	-9.65	.0	-9.67	.0	-9.60	.0	-9.47	.0
	-3.77	.0	-3.91	.0	-4.14	.0	-4.41	.0
	-.002	-.342	-.003	-.338	.019	-.351	.053	-.376
	-.002	.342	-.003	.338	.019	.351	.053	.376
	-.338	-1.45	-.456	-1.98	-.541	-2.44	-.595	-2.84
	-.338	1.45	-.456	1.98	-.541	2.44	-.595	2.84
	-.038	.0	-.034	.0	-.034	.0	-.016	.0
	-.598	.0	-.728	.0	-.726	.0	-.685	.0
Closed loop poles	-9.59	.0	-9.67	.0	-9.64	.0	-9.53	.0
	-3.47	.0	-3.93	.0	-4.29	.0	-4.63	.0
	-1.89	-1.21	-1.86	-1.96	-1.91	-2.50	-1.94	-2.94
	-1.89	1.21	-1.86	1.96	-1.91	2.50	-1.94	2.94
	-.124	.0	-.169	.0	-.414	-.365	-.365	-.445
	-.394	.357	-.413	-.334	-.414	.365	-.365	.445
	-.394	-.357	-.413	.334	-.195	.0	-.202	.0
	-.522	.0	-.633	.0	-.655	.0	-.696	.0
Desired poles	-9.65	.0	-9.67	.0	-9.60	.0	-9.47	.0
	-3.77	.0	-3.91	.0	-4.14	.0	-4.41	.0
	-2.00	-1.45	-2.00	-2.00	-2.00	-2.44	-2.00	-2.84
	-2.00	1.45	-2.00	2.00	-2.00	2.44	-2.00	2.84
	-.400	-.342	-.400	-.338	-.400	-.351	-.400	-.376
	-.400	.342	-.400	.338	-.400	.351	-.400	.376
	-.598	.0	-.728	.0	-.726	.0	-.685	.0
	-.200	.0	-.200	.0	-.200	.0	-.200	.0

	q	Δθ	p	ΔΦ	r
u_1 Longitudinal cyclic control	.876	.016	-.813	-1.08	-.780
u_3 Lateral cyclic control	2.52	3.78	.336	2.84	-1.57
u_4 Pedal	2.39	2.19	.114	-.265	1.19

Parameters of controller
see fig.3.10.

Table 3.2: Poles of feedback system and parameters of controller.

Controller design

First one determines for which poles the difference between the desired and open-loop poles are largest. Then the input is selected for which the controllability coefficients for the above determined poles is largest. In table 3.3 the poles and belonging controllability-coefficients are shown. For shifting the "fast" conjugate-complex pole-pair one has to choose input u_4. Two further design steps with input u_1 for shifting the "slow" conjugate-complex pole-pair and input u_3 for shifting the "slow" real poles one gets the result of table 3.2. No further improvement could be found by using input u_2.

	a)		b)			
	poles		controllability-coefficients			
	Re	Im	u_1	u_2	u_3	u_4
40 knots	-9.658	0.000	0.0380	0.0138	5.6380	0.8400
	-3.770	0.000	1.0340	0.0218	0.0025	0.0003
	-0.338	-1.458	0.0042	0.0839	0.0047	1.6540
	-0.338	1.458	0.0042	0.0839	0.0047	1.6540
	0.009	0.334	0.0610	0.0024	0.0004	0.0005
	0.009	-0.334	0.0610	0.0024	0.0004	0.0005
	-0.074	0.000	0.0052	0.0064	0.0470	0.0247
	-0.586	0.000	0.5264	0.0752	0.0000	0.0010
60 knots	-9.673	0.000	0.0405	0.0067	5.6000	0.9510
	-3.916	0.000	1.0720	0.0827	0.0051	0.0008
	-0.456	1.988	0.0085	0.0173	0.0016	2.1020
	-0.456	-1.988	0.0085	0.0173	0.0016	2.1020
	0.002	0.338	0.0666	0.0052	0.0002	0.0003
	0.002	-0.338	0.0666	0.0052	0.0002	0.0003
	-0.033	0.000	0.0033	0.0025	0.0477	0.0207
	-0.729	0.000	0.5753	0.0028	0.0000	0.0019
80 knots	-9.600	0.000	0.0427	0.0121	5.5700	1.0460
	-4.143	0.000	1.1410	0.1827	0.0111	0.0016
	-0.541	2.440	0.0093	0.0606	0.0026	2.5560
	-0.541	-2.440	0.0093	0.0606	0.0026	2.5560
	0.018	0.356	0.0805	0.0112	0.0002	0.0003
	0.018	-0.356	0.0805	0.0112	0.0002	0.0003
	-0.032	0.000	0.0033	0.0000	0.0487	0.0207
	-0.726	0.000	0.5308	0.0079	0.0000	0.0019
100 knots	-9.479	0.000	0.0439	0.0188	5.5370	1.0280
	-4.414	0.000	1.2340	0.3217	0.0237	0.0023
	-0.595	2.844	0.0045	0.0875	0.0050	2.9100
	-0.595	-2.844	0.0045	0.0875	0.0050	2.9100
	0.053	0.376	0.1005	0.0215	0.0003	0.0004
	0.053	-0.376	0.1005	0.0215	0.0003	0.0004
	-0.017	0.000	0.0037	0.0056	0.0537	0.0264
	-0.685	0.000	0.4261	0.0294	0.0002	0.0016

Table 3.3: controllability coefficients.

Results

In table 3.2 the open-loop, achieved closed loop and desired poles are shown. One can see, that only minor differences remain for all four cases.
For the 80 knots case the different pole locations are shown in figures 3.13 a,b.

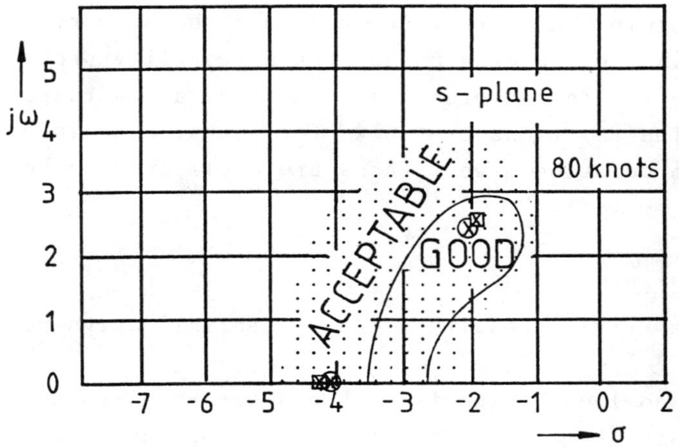

Fig.3.13a: Desired and attained poles; "fast"-area.

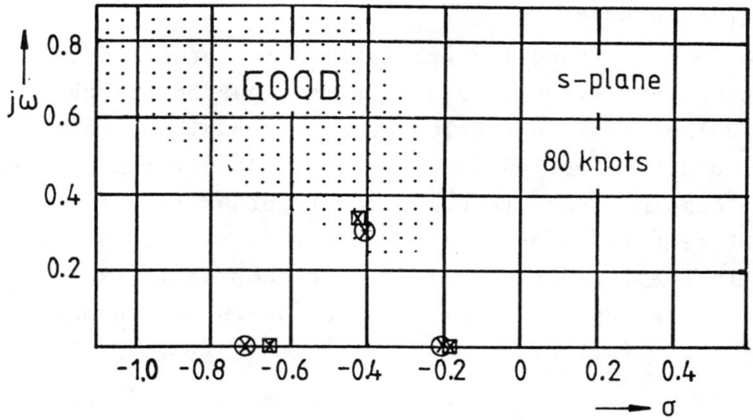

Fig.3.13b: Desired and attained poles; "slow"-area.

3.1.10 SENSOR - AND ACTUATOR - FAILURE

In a multivariable plant with no redundant inputs an actuator failure causes the loss of independent control of the outputs. Although the feedback system could be still stable, the plant often is shut down in these cases, because the overall system behavior becomes too poor. In some systems a shut down is not allowed, suppose vehicles, aircrafts, helicopters etc. In this systems the overall function must be preserved despite failures. In the sequel a method for a stability-preserving design is given in the presence of sensor- and actuator-failures. The relations are shown for actuator-failures, but with little additional work those are applicable to sensor-failures.

Actuator-failure-design

First a trivial result of the sequential design method is given:
If the closed-loop-system is stable after the i-th design recursion step (that is after the design of the controller for input i) then the total closed loop system remains stable, when all the actuators j fail with $i < j \leq r$. r is the number of plant inputs.
Now the design method for arbitrary actuator failure is given. Each actuator failure combination is regarded together with the actual plant as a single new plant model and then the recursive design method is applicated. At first all possible failure combinations are collected together in a table. Then the different failure-combinations are arranged to failure-groups in the following way: failure-combination k belongs to the failure-group j, iff all the actuators with number i, $j \leq i \leq r$ are failed.

If all possible failure-combinations are allowed, one gets the arranged table 3.4 immediately. The status of actuator 1 beginning with status 0 (failed) changes from row to row. The status of actuator 2 beginning with status 0 changes each second row and in general the status of actuator i beginning with status 0 changes the status in each 2^{i-1}-th row. The whole number of combinations is (2^r-1) after deleting the first row (the uncontrolled plant). r is the number of inputs. To each failure group j there belong 2^{j-1} variation cases. The failure-group j belongs to the input i = j. An example is given with four inputs in table 3.4.

actuator				group
1	2	3	4	j
0	0	0	0	1
1	1	0	0	2
0	1	0	0	
1	0	1	0	3
0	0	1	0	
1	1	1	0	
0	1	1	0	
1	0	0	1	4
0	0	0	1	
1	1	0	1	
0	1	0	1	
1	0	1	1	
0	0	1	1	
1	1	1	1	
0	1	1	1	

Table 3.4: Arrangement of failure-combinations, failed ≡ 0 and working ≡ 1.

The recursive controller-design starts now with failure group 1. Before arranging the failure-combination-table the sequence of the inputs is not defined. There are at most r! different possible sequences. An aid to find a proper sequence are the controllability coefficients of section 3.1.8 .

If all failure combinations are permitted, then it is necessary for finding a stabilizing controller, that all unstable modes are controllable from each input.

If there are simultaneous actuator failures and different variation-cases of the plant, then only the number of total variationcases is increased. Is m the number of plant-variation cases and is l the number of different failure combinations, then one has to consider ml separate models.

Example:

Given is an unstable plant with three inputs and up to two actuators may fail at the same time. The arranged failure-group table is shown in the following table.

actuator			failure-group
u_1	u_2	u_3	
0	0	0	
1	0	0	1
0	1	0	2
1	1	0	
0	0	1	3
1	0	1	
0	1	1	
1	1	1	

Under the assumption, that there are no parameter-variations in the original plant, the following figures 3.14 show the sequential controller-design.

a)

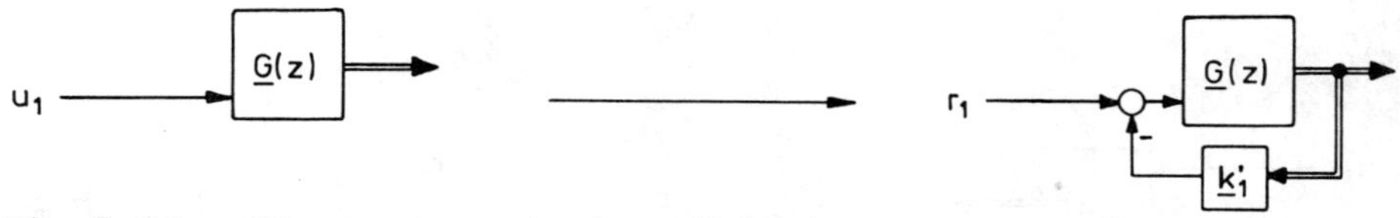

Fig.3.14a: First step, design of $\underline{k}_1$.

b)

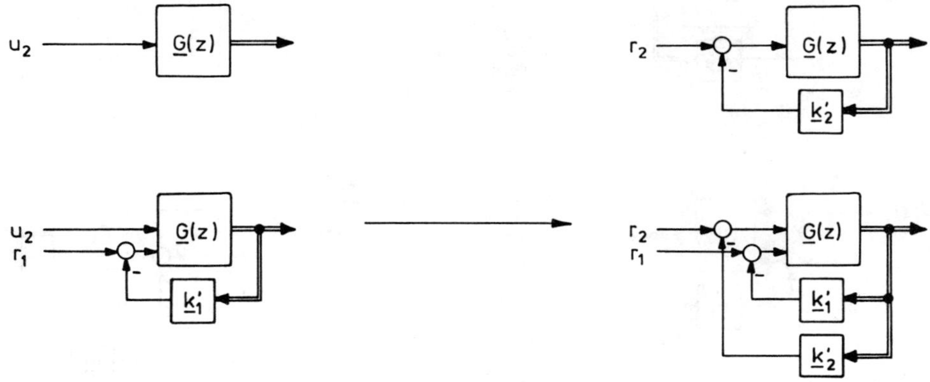

Fig.3.14b: Second step, $\underline{k}_2$ is determined in a multimodel design.

c)

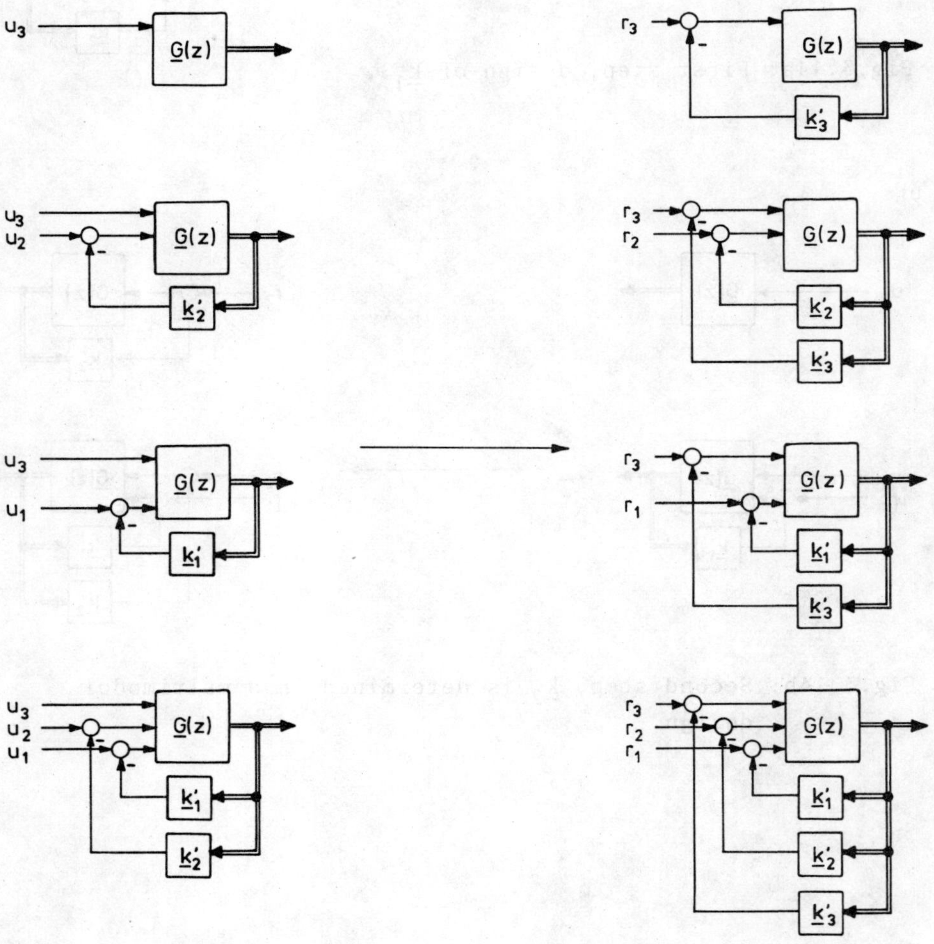

Fig.3.14c: $\underline{k}_3$ is determined like second step and the same for all four closed-loop circuits.

3.2 INPUT - METHOD

3.2.1 INTRODUCTION

Suppose the best closed-loop response under the influence of a well-known disturbance or control-input is given for each model-variation case. Now we look for a controller, which is the same for all model-variation cases and minimizes the difference between the desired and attained closed-loop response. In the sequel a method for finding such a controller is described, further information see Lange [6].
Assume now the desired plant-input (i.e. controller-output) for each variation case is given for special control-inputs or disturbance, compare figure 3.15.

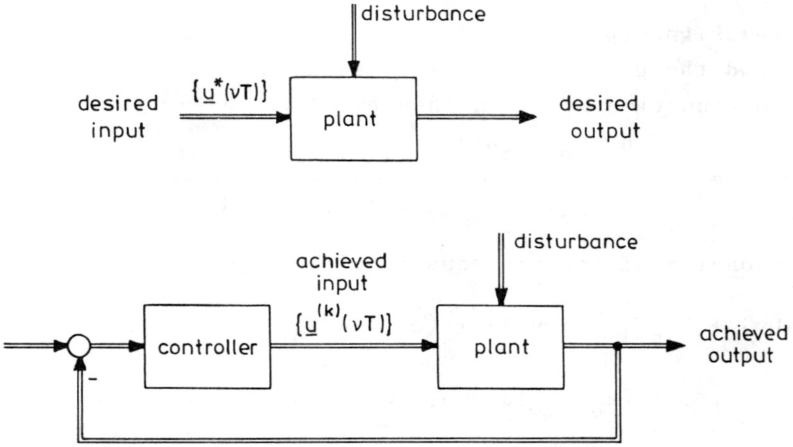

Fig.3.15: Inputs and outputs of plant for open and closed loop system.

If there are l plant-variation cases, the controller-design objective in the discrete-time-case is

$$\underset{\text{controller-parameter}}{\text{Min}} \sum_{k=1}^{l} \sum_{\nu=o}^{\infty} \left(u_i^{(k)}(\nu T) - u_i^{*}(\nu T) \right)^2 \quad \text{for each } i = 1 \ldots r$$

T sampling period.

This can be interpreted as a deterministic identifikation problem for finding the controller transfer-function.

3.2.2 FINDING A CONTROLLER WITH A PREASSIGNED POLE AND A PREASSIGNED ZERO

First the design relations are given, if only one controller-input $\{e(\nu)\}$, one controller-output $\{u(\nu)\}$ and one plant variation case must be considered, compare the figure 3.16. The problem is to find the parameters of the linear, timeinvariant controller transfer function $G(z)$.

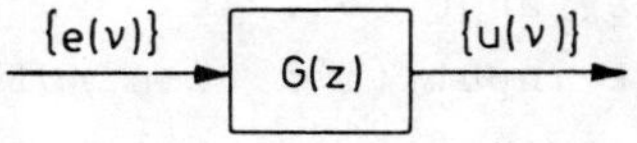

Fig.3.16: controller in- and output.

The preassigned pole (for instance an integrator) lies at $z = \alpha$ and the preassigned zero lies at $z = \beta$. The controller transfer-function is given then by

$$G(z) = \frac{(b_n z^n + b_{n-1} z^{n-1} + \dots + b_o)}{(z^n + a_{n-1} z^{n-1} + \dots + a_o)} \; \frac{(z-\beta)}{(z-\alpha)} = \frac{u(z)}{e(z)} \quad .$$

The belonging difference equation is

$$u(\nu) + (a_{n-1}-\alpha)u(\nu-1) + (a_{n-2}-\alpha\, a_{n-1}) \quad u(\nu-2) + \dots$$

$$\dots + (a_o-\alpha a_1)u(\nu-n) - \alpha \; a_o \; u(\nu-n-1) \quad =$$

$$b_n e(\nu) + (b_{n-1}-\beta b_n)e(\nu-1) + (b_{n-2}-\beta b_{n-1})e(\nu-2) + \dots$$

$$\dots + (b_o-\beta b_1)e(\nu-n) - \beta b_o e(\nu-n-1)$$

With abbreviations

$$\tilde{u}(\nu) = u(\nu) - \alpha \quad u(\nu-1)$$

$$\tilde{e}(\nu) = e(\nu) - \beta \quad e(\nu-1)$$

we get

$$\tilde{u}(\nu) + a_{n-1}\tilde{u}(\nu-1) + \dots + a_o\tilde{u}(\nu-n) = b_n\tilde{e}(\nu) + \dots + b_o\tilde{e}(\nu-n).$$

The $a_i's$ and $b_i's$ are components of the parametervector

$$\underline{\theta}' = [-a_{n-1},\ldots,-a_o,\ b_n,\ldots,b_o]$$

and the input and output data are components of the data-vector

$$\underline{\phi}(\nu) = \begin{bmatrix} u(\nu-1) \\ \vdots \\ u(\nu-n) \\ e(\nu) \\ \vdots \\ e(\nu-n) \end{bmatrix} - \begin{bmatrix} \alpha u(\nu-2) \\ \vdots \\ \alpha u(\nu-n-1) \\ \beta\ e(\nu-1) \\ \vdots \\ \beta\ e(\nu-n-1) \end{bmatrix} = \begin{bmatrix} \tilde{u}(\nu-1) \\ \vdots \\ \tilde{u}(\nu-n) \\ \tilde{e}(\nu) \\ \vdots \\ \tilde{e}(\nu-n) \end{bmatrix} .$$

The output data $\tilde{u}(\nu)$ then becomes

$$\tilde{u}(\nu) = \underline{\theta}'\underline{\phi}(\nu) .$$

There may remain a difference $\varepsilon(\nu)$ between the actual output $\tilde{u}(\nu)$ and the output determined with the parameter vector $\underline{\theta}$ and the data-vector $\underline{\phi}(\nu)$:

$$\tilde{u}(\nu) - \underline{\theta}'\underline{\phi}(\nu) = \varepsilon(\nu) .$$

This difference $\varepsilon(\nu)$ can be weighted depending on ν

$$\gamma(\nu)(\tilde{u}(\nu) - \underline{\theta}'\underline{\phi}(\nu)) = \gamma(\nu)\varepsilon(\nu) .$$

DETERMINATION OF $\underline{\theta}$

With a sufficient long sequence of output data $\{\tilde{u}(\nu)\}$ and data-vectors $\underline{\phi}(\nu)$ the parameter-vector $\underline{\theta}$ is determined with the help of the least squares method. The differences are weighted with the matrix $\underline{\Gamma}$:

$$\underline{\Gamma} = \begin{bmatrix} \gamma^{(1)} & & & \\ & \gamma^{(2)} & & \underline{0} \\ & & \ddots & \\ \underline{0} & & & \gamma^{(m)} \end{bmatrix} .$$

With
$$\underline{u}' = [\tilde{u}(1)\ ,\ \ldots\ldots,\ \tilde{u}(m)]$$
$$\underline{\Phi} = [\underline{\psi}(1)\ ,\ \ldots\ldots,\ \underline{\psi}(m)]$$

one gets

$$\underline{\theta}' = \underline{u}'\ \underline{\Gamma}\ \underline{\Gamma}'\underline{\Phi}'[\underline{\Phi}\ \underline{\Gamma}\ \underline{\Gamma}'\underline{\Phi}']^{-1}\ .$$

3.2.3 MORE THAN ONE VARIATION CASES

We start with

$$\underline{u}_k' = [\tilde{u}(1),\ \ldots\ldots,\ \tilde{u}(m)]_k$$
$$\underline{\Phi}_k = [\psi(1),\ \ldots\ldots,\ \underline{\psi}(m)]_k$$

k is the variation-case index.

By lining up the sequence of outputvariables and the sequence of vectors $\underline{\psi}$ for the variation-cases 1 to l we get

$$\underline{U}' = \left[\underline{u}_1,\ \ldots\ldots,\ \underline{u}_l\right]$$
$$\underline{H} = \left[\underline{\Phi}_1,\ \ldots\ldots,\ \underline{\Phi}_l\right]$$

In the same way we get the new weighting-matrix

$$\hat{\underline{\Gamma}} = \begin{bmatrix} \underline{\Gamma}_1 & & \underline{O} \\ & \ddots & \\ \underline{O} & & \underline{\Gamma}_l \end{bmatrix}$$

and with the least-squares method the result for the parameter-vector is

$$\underline{\theta}' = \underline{U}'\hat{\underline{\Gamma}}\ \hat{\underline{\Gamma}}'\underline{H}'\left[\underline{H}\ \hat{\underline{\Gamma}}\ \hat{\underline{\Gamma}}'\underline{H}'\right]^{-1}\ .$$

3.2.4 MORE THAN ONE CONTROLLER-INPUTS

In the multivariable case the controller for one input has the following structure

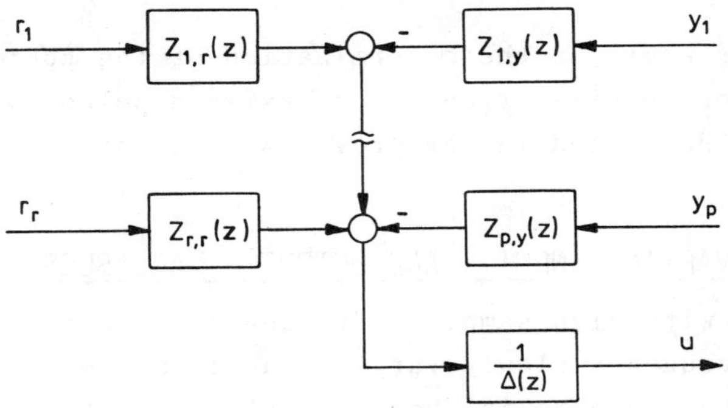

Fig.3.17: Multivariable case, controller for one plant-input.

The Z(z), N(z) are polynomials in z with degree Z(z) $\leq$ degree N(z). This polynomials are written in the following way:

$$Z_{i,r}(z) = \underline{b}_{i,r}' \begin{bmatrix} z^n \\ \vdots \\ z \\ 1 \end{bmatrix}, \quad Z_{i,y}(z) = \underline{b}_{i,y}' \begin{bmatrix} z^n \\ \vdots \\ z \\ 1 \end{bmatrix}, \quad N(z) = \underline{a}' \begin{bmatrix} z^{n-1} \\ \vdots \\ z \\ 1 \end{bmatrix}$$

Using this representation we get for the controller-output:

$$u(\nu) = \left[-\underline{a}', \underline{b}_{1,r}', \ldots, \underline{b}_{r,r}', \underline{b}_{1,y}', \ldots, \underline{b}_{p,y}' \right] \begin{bmatrix} \underline{u}(\nu) \\ \underline{r}_1(\nu) \\ \vdots \\ \underline{r}_r(\nu) \\ \underline{y}_1(\nu) \\ \vdots \\ \underline{y}_p(\nu) \end{bmatrix}$$

with

$$\underline{u}(\nu) = \begin{bmatrix} u(\nu-1) \\ \vdots \\ u(\nu-n) \end{bmatrix} \; ; \quad \underline{r}_i(\nu) = \begin{bmatrix} r_i(\nu) \\ \vdots \\ r_i(\nu-n) \end{bmatrix} \; ; \quad \underline{y}_i(\nu) = \begin{bmatrix} y_i(\nu) \\ \vdots \\ y_i(\nu-n) \end{bmatrix} .$$

The further procedure in the determination of the controller-parameters under consideration of preassigned poles and zeros is the same as described in the previous sections.

3.2.5 SLOWLY VARYING INPUT - AND OUTPUT - VARIABLES

In systems with high sampling frequencies and relative to the sampling frequency slowly varying input- and output-variables a lot of the elements in the $\underline{r}(\nu)$, $\underline{y}(\nu)$ and $\underline{u}(\nu)$ vectors have the same values. Out of this the following question arises:
Could'nt it be possible to sample less frequently (or to use less samples), determine out of this samples the transfer-functions of the slowly sampled system and then transform this slowly-sampled system into an equivalent quick sampled system?

This is possible, if one proceeds in the following way:

First the parameters of the slowly sampled system are determined

$$\underline{x}(\nu+1) = \underline{\Phi}_T \underline{x}(\nu) + \underline{H}_T \begin{bmatrix} \underline{r}(\nu) \\ \\ \underline{y}(\nu) \end{bmatrix}$$

$$\underline{u}(\nu) = \underline{C}\; \underline{x}(\nu) + \underline{D} \begin{bmatrix} \underline{r}(\nu) \\ \\ \underline{y}(\nu) \end{bmatrix}$$

T is the index of matrices for the low sampling-frequency.

Then an equivalent, step-invariant continuous-time system is determined :

$$\dot{\underline{z}}(t) = \underline{A}\; \underline{z}(t) + \underline{B}\; \underline{v}(t)$$

$$\underline{w}(t) = \underline{C}\; \underline{z}(t) + \underline{D}\; \underline{v}(t) .$$

This system is found by re-transforming a discret-time system into a continous-time-system. The re-transformation is done with the following relations:

$$\underline{A} = \frac{1}{T} \ln(\underline{\Phi}_T) \;, \quad T \text{ is the sampling-period}$$

and

$$\underline{B} = \left[\int_0^T e^{\underline{A}\tau} \, d\tau \right]^{-1} \underline{H}_T \;;$$

$\underline{A}$ and $\underline{B}$ are determined with the help of the matrix-functions out of [4].

If $\underline{\Phi}_T$ is diagonal-equivalent with eigenvalues $\lambda_1,\dots,\lambda_n$, then $\underline{A}$ and $\underline{B}$ are determined in the following way:
$\underline{\Phi}_T = \underline{T}^{-1}\text{diag}(\lambda_1,\dots\lambda_n)\ \underline{T}$; $\underline{T}$ is the eigenvector-matrix and with

$$\underline{T}\ \underline{\Phi}_T \underline{T}^{-1} = e^{[\underline{T}(\underline{A}\ T)\underline{T}^{-1}]}$$

one gets

$$\underline{A} = \underline{T}^{-1}\text{diag}(\alpha_1,\dots,\alpha_n)\underline{T}$$

with

$$\alpha_i = \frac{\ln \lambda_i}{T} \quad .$$

The ln-function is multi-valued if there is a complex argument. Under this condition one has to take the value with the smallest absolut amount of the imaginary part.

The input-matrix becomes

$$\underline{B} = \underline{T}^{-1}\text{diag}[\beta_1,\dots,\beta_n]\ \underline{T}\ \underline{H}_T$$

with

$$\beta_i^{-1} = \begin{cases} \dfrac{T}{\ln\lambda_i}\left[e^{\ln\lambda_i} - 1\right] & \text{for} \quad \lambda_i \neq 1 \\ T & \text{for} \quad \lambda_i = 1 \quad . \end{cases}$$

The new discret-time system then is found by sampling the continous-time system with sampling period ϑ.

We get

$$\underline{\Phi}_\vartheta = e^{\underline{A}\vartheta} = \underline{T}^{-1}\text{diag}[\gamma_1,\ldots,\gamma_n]\ \underline{T}$$

with

$$\gamma_i = \lambda_i^{\frac{\vartheta}{T}} .$$

The new poles are lying on the "constant-damping-curves" for discret-time systems.

The new input-matrix becomes

$$\underline{H}_\vartheta = \underline{T}^{-1}\text{diag}[\varphi_1,\ldots,\varphi_n]\ \underline{T}\ \underline{H}_T$$

with

$$\varphi_i = \begin{cases} \dfrac{\lambda_i^{\frac{\vartheta}{T}}-1}{\lambda_i-1} & \text{for } \lambda_i \neq 1 \\[2ex] \dfrac{\vartheta}{T} & \text{for } \lambda_i = 1 \end{cases} .$$

3.2.6 ADAPTING THE CONTROLLER TO STEP-WISE VARYING SAMPLING-TIME

If there are step-wise varying sampling-times in the closed-loop-system, then it is easy to adapt the controller to the new sampling-time with help of the procedure, described in the previous section. If the controller has to remain step-invariant all the mathematics of the previous section is used without any change. For quick calculation the matrices $\underline{T}$, $\underline{T}^{-1}$ and the controller-eigenvalues λ_i should be tabulated.

Closed-loop stability in the presence of step-wise varying sampling-time.

a) The controller is adapted to the different sampling-times. The closed-loop system has the structure of figure 3.18.

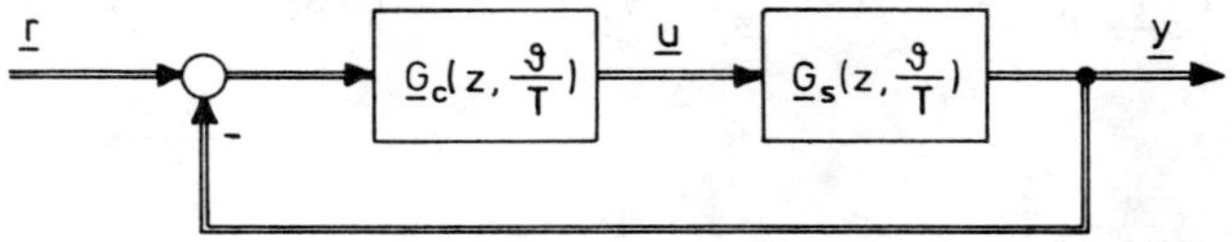

Fig.3.18: Closed loop circuit, depending on variable sampling time.

It is supposed that $\underline{\Phi}_c$ is diagonal-equivalent and has eigenvalues $\lambda_{1,c},\dots,\lambda_{m,c}$.
The controller-transfer function for $\vartheta = T$ then is

$$\underline{G}_c(z,1) = \underline{C}_c(z\underline{E}-\underline{\Phi}_c)^{-1}\underline{H}_c + \underline{D}_c$$
$$= \underline{C}_c\underline{T}_c^{-1}[z\underline{E}-\text{diag}(\lambda_{1,c}\dots\lambda_{m,c})]^{-1}\underline{T}_c\underline{H}_c + \underline{D}_c .$$

With $\underline{\Phi}_s$ diagonal-equivalent with eigenvalues $\lambda_{1,s}\dots\lambda_{n,s}$ we get for $\vartheta = T$

$$\underline{G}_s(z,1) = \underline{C}_s(z\underline{E}-\underline{\Phi}_s)^{-1}\underline{H}_s + \underline{D}_s$$
$$= \underline{C}_s\underline{T}_s^{-1}[z\underline{E}-\text{diag}(\lambda_{1,s}\dots\lambda_{n,s})]^{-1}\underline{T}_s\underline{H}_s + \underline{D}_s .$$

Now there occurs a step-wise change in sampling time. With the relations of the last sections one gets

$$\underline{G}_c(z,\tfrac{\vartheta}{T}) = \underline{C}_c\underline{T}_c^{-1}[z\underline{E}-\text{diag}(\gamma_{1,c}\dots\gamma_{m,c})]^{-1}\,\text{diag}(\varphi_{1,c}\dots\varphi_{m,c})\cdot$$
$$\cdot\ \underline{T}_c\underline{H}_c + \underline{D}_c$$

$$\gamma_{i,c} = \lambda_{i,c}^{\frac{\vartheta}{T}}$$

$$\varphi_{i,c} = \begin{cases} \dfrac{\lambda_{i,c}^{\frac{\vartheta}{T}}-1}{\lambda_{i,c}-1} & \text{for } \lambda_{i,c} \neq 1 \\[2ex] \dfrac{\vartheta}{T} & \text{for } \lambda_{i,c} = 1 \end{cases}$$

and

$$\underline{G}_s(z,\tfrac{\vartheta}{T}) = \underline{C}_s\underline{T}_s^{-1}[z\underline{E}-\mathrm{diag}(\gamma_{1,s}\cdots\gamma_{n,s})]^{-1}\ \mathrm{diag}(\varphi_{1,s}\cdots\varphi_{n,s})\cdot$$

$$\cdot\ \underline{T}_s\underline{H}_s + \underline{D}_s$$

$$\gamma_{i,s} = \lambda_{i,s}^{\frac{\vartheta}{T}}$$

$$\varphi_{i,s} = \begin{cases} \dfrac{\lambda_{i,s}^{\frac{\vartheta}{T}}-1}{\lambda_{i,s}-1} & \text{for} \quad \lambda_{i,s} \neq 1 \\ \dfrac{\vartheta}{T} & \text{for} \quad \lambda_{i,s} = 1 \ . \end{cases}$$

Nowe the closed loop dynamic-matrix $\underline{\Phi}_{CL}(\frac{\vartheta}{T})$ is easily determined.
This matrix depends only on $\frac{\vartheta}{T}$. By inspection of the eigenvalues of $\underline{\Phi}_{CL}(\frac{\vartheta}{T})$ for different $\frac{\vartheta}{T}$ the stability of the closed loop system with respect to varying sampling time can be examined.

b) The controller is not adapted to the different sampling-times.
The closed-loop dynamic matrix $\underline{\Phi}_{CL}(\frac{\vartheta}{T})$ is now determined with the elements of $\underline{G}_c(z,1)$ and $\underline{G}_s(z,\frac{\vartheta}{T})$.

3.3 ROBUST STABILITY

3.3.1 INTRODUCTION

In this section necessary and sufficient conditions for robust stability of the closed-loop system are given. The stability is tested with help of the principal gains. It is easy to applicate the stability criteria because one first plots the frequency depending principal gains and then checks, if one of this curves touchs or cuts a part of the negativ real axis. Here the well-known "critical point" becomes a "critical part of the real axis". The criteria are formulated for continuous systems. With redefinition of the Nyquist-contour or after bilinear transformation the criteria are applicable

to discret-time systems without any change.
The represented stability criteria are an extension of the criteria given in [5], further information see Lange [6].

Definition: The principal gains of the transfer-matrix $\underline{F}(s)$ are the locuses of the eigenvalues of $\underline{F}(s)$ in the complex plane depending on the complex frequency-variable s of the Nyquist-contour. The Nyquist-contour is

a) for stability-tests of continuous-time systems the rim of the right-half-s-plane,

b) for stability-tests of discret-time systems the rim of the unit circle in the belonging complex plane.

3.3.2 MULTIPLICATIVE PERTURBATIONS

Given is the closed loop structure of the next figure.

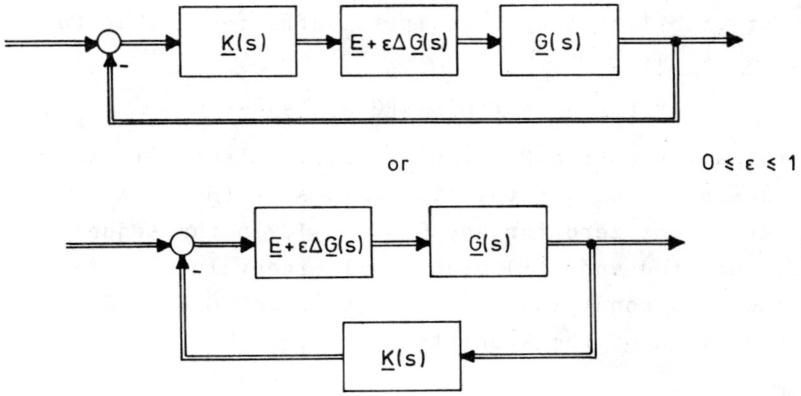

Fig.3.19: Closed-loop system.

$\underline{K}(s)$ is the controller, $\underline{G}(s)$ is the unperturbated plant, $\Delta\underline{G}(s)$ is the perturbation transferfunction, ε is a factor, varying between $\varepsilon = 0$ and $\varepsilon = 1$.

THEOREM 1

Under the conditions

- multiplicative perturbation model
  $\underline{\tilde{G}}(s) = (\underline{E} + \varepsilon\Delta\underline{G}(s))\,\underline{G}(s)$ with $0 \leq \varepsilon \leq 1$
- $\Delta\underline{G}(s)$ bibostable
- $\underline{T}(s) = \underline{G}(s)\underline{K}(s)\,[\underline{E} + \underline{G}(s)\underline{K}(s)]^{-1}$ bibostable

the closed loop system of figure 3.19 is bibostable, iff none of the principal gains of $\Delta\underline{G}(s)\,\underline{T}(s)$ touchs or cuts the real axis of the principal-gain-plane in the interval $(-\infty,-1]$.

Proof

a) Sufficient condition

We start with a stable $\underline{T}(s) = \underline{G}(s)\underline{K}(s)[\underline{E}+\underline{G}(s)\underline{K}(s)]^{-1}$.
For destabilizing this system, at least for one ε

$$\det(\underline{E} + (\underline{E} + \varepsilon\Delta\underline{G}(s))\underline{G}(s)\underline{K}(s))$$

must become zero.
For stability therefore it is sufficient, that for all $0 \leq \varepsilon \leq 1$

$$\det(\underline{E} + (\underline{E} + \varepsilon\Delta\underline{G}(s))(\underline{G}(s)\underline{K}(s)) \neq 0$$

for all s of the Nyquist-contour. This condition is the same as: none of the eigenvalues of $(\underline{E}+(\underline{E}+\varepsilon\Delta\underline{G}(s))\underline{G}(s)\underline{K}(s))$ may become zero for any $0 \leq \varepsilon \leq 1$. In the sequel the (s) is omitted and $\lambda(\cdot)$ means: all eigenvalues of $(\cdot)$.
The last condition becomes $\lambda(\underline{E}+(\underline{E}+\varepsilon\Delta\underline{G})\underline{G}\,\underline{K}) \neq 0$ for $0\leq\varepsilon\leq 1$ and all s of the Nyquist-contour.

With

$$\underline{E} + (\underline{E} + \varepsilon\,\Delta\underline{G})\underline{G}\,\underline{K} = \underline{E} + \underline{G}\,\underline{K} + \varepsilon\Delta\underline{G}\,\underline{G}\,\underline{K}$$

we get

$$\lambda\{[(\underline{E} + \varepsilon\Delta\underline{G}\,\underline{G}\,\underline{K}(\underline{E} + \underline{G}\,\underline{K})^{-1}](\underline{E} + \underline{G}\,\underline{K})\} \neq 0$$

for $0 \leq \varepsilon \leq 1$ and all s of the Nyquist-contour.

$\lambda(\underline{E} + \underline{G}\ \underline{K}) \neq 0$ by prerequisite. The rank of the product of a regular matrix with a irregular matrix is equal to the rank of the irregular matrix and we get the condition $\lambda(\underline{E} + \varepsilon\Delta\underline{G}\ \underline{G}\ \underline{K}(\underline{E} + \underline{G}\ \underline{K})^{-1}) \neq 0$ for $0 \leq \varepsilon \leq 1$ and all s of the Nyquist-contour.

Because there is fulfiled

$$\lambda(\Delta\underline{G}\ \underline{T}) \neq -\varepsilon^{-1}$$

for $0 \leq \varepsilon \leq 1$ and all s of the Nyquist-contour the feedback system fig.3.19 is bibostable.

b) Necessary condition

$$\underline{\tilde{T}} = (\underline{E} + \varepsilon\Delta\underline{G})\underline{G}\ \underline{K}(\underline{E} + (\underline{E} + \varepsilon\Delta\underline{G})\underline{G}\ \underline{K})^{-1}$$

is supposed to be bibostable.

We get

$$\begin{aligned}\underline{\tilde{T}} &= (\underline{E} + \varepsilon\Delta\underline{G})\underline{G}\ \underline{K}(\underline{E} + \underline{G}\ \underline{K} + \varepsilon\Delta\underline{G}\ \underline{G}\ \underline{K})^{-1}\\ &= (\underline{E} + \varepsilon\Delta\underline{G})\underline{G}\ \underline{K}[(\underline{E} + \varepsilon\Delta\underline{G}\ \underline{G}\ \underline{K}(\underline{E} + \underline{G}\ \underline{K})^{-1})(\underline{E} + \underline{G}\ \underline{K})]^{-1}\\ &= (\underline{E} + \varepsilon\Delta\underline{G})\underline{G}\ \underline{K}(\underline{E} + \underline{G}\ \underline{K})^{-1}[\underline{E} + \varepsilon\Delta\underline{G}\ \underline{G}\ \underline{K}(\underline{E} + \underline{G}\ \underline{K})^{-1}]^{-1}\end{aligned}$$

$$\begin{aligned}\underline{\tilde{T}} &= (\underline{E}+\varepsilon\Delta\underline{G})\underline{G}\ \underline{K}(\underline{E}+\underline{G}\ \underline{K})^{-1}[\underline{E}+\varepsilon\Delta\underline{G}\ \underline{G}\ \underline{K}(\underline{E}+\underline{G}\ \underline{K})^{-1}]^{-1}\\ &= [\underbrace{\underline{G}\ \underline{K}(\underline{E}+\underline{G}\ \underline{K})^{-1}}_{\underline{T}}+\underbrace{\varepsilon\Delta\underline{G}\ \underline{G}\ \underline{K}(\underline{E}+\underline{G}\ \underline{K})^{-1}}_{\varepsilon\underline{W}}][\underline{E}+\underbrace{\varepsilon\Delta\underline{G}\ \underline{G}\ \underline{K}(\underline{E}+\underline{G}\ \underline{K})^{-1}}_{\varepsilon\underline{W}}]^{-1}\\ &= [\underline{T} + \varepsilon\underline{W}][\underline{E} + \varepsilon\underline{W}]^{-1}\end{aligned}$$

$$\tilde{T} = \underline{T}(E + \varepsilon\underline{W})^{-1} + \varepsilon\underline{W}(\underline{E} + \varepsilon\underline{W})^{-1}$$

$\underline{\tilde{T}}$ is supposed to be bibostable, therefore $(\underline{E} + \varepsilon\underline{W})^{-1}$ must be bibostable for $0 \leq \varepsilon \leq 1$. $(\underline{E} + \varepsilon\underline{W})^{-1}$ is bibostable only, if all eigenvalues of $\underline{W} = \Delta\underline{G}\ \underline{T}$ take on values out of the interval $(-\infty,-1]$ for all s of the Nyquist-contour.

SINGULAR VALUES [5]

If $\bar{\sigma}(\underline{T}) < \dfrac{1}{\bar{\sigma}[\Delta\underline{G}]}$

then the above given stability-condition is also fulfiled.

$\bar{\sigma}(\cdot)$ are the maximum singular values over all s of the Nyquist-contour of the matrix $(\cdot)$.

In the next figure the principal-gain stability-test and singular-values-stability-test are compared.

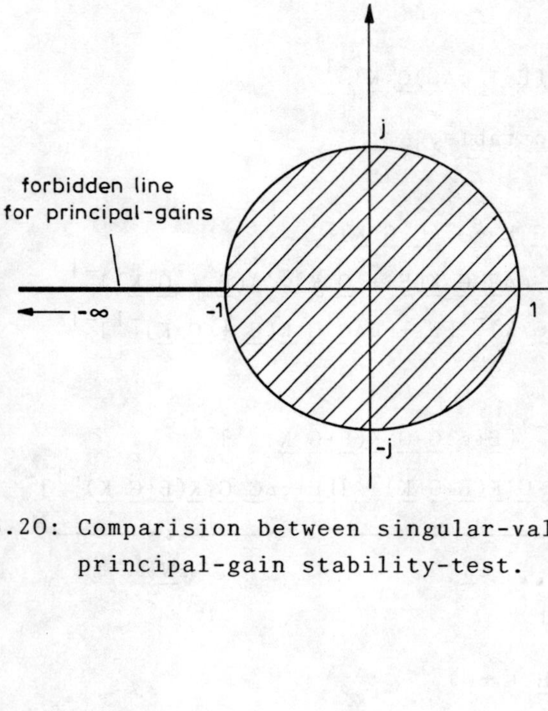

Fig.3.20: Comparision between singular-value and principal-gain stability-test.

3.3.3 ADDITIV PERTURBATIONS

Given is the closed-loop circuit of the following figure.

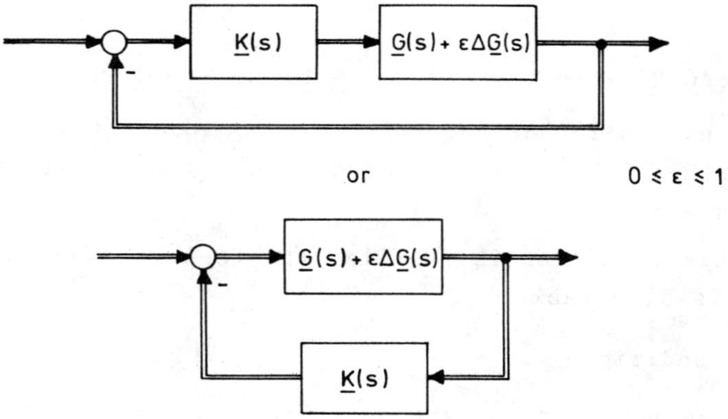

Fig.3.21: Closed-loop system with additiv perturbations.

THEOREM 2:

Under the conditions

- additiv perturbation model $\tilde{\underline{G}}(s) = \underline{G}(s) + \varepsilon\Delta\underline{G}(s)$ with $0 \leq \varepsilon \leq 1$
- $\Delta\underline{G}(s)$ bibostable
- $\underline{T}(s) = \underline{G}(s)\underline{K}(s)[\underline{E} + \underline{G}(s)\underline{K}(s)]^{-1}$ bibostable

the closed loop system of figure 3.21 is bibostable, iff none of the principal gains of $\Delta\underline{G}(s)\underline{K}(s)[\underline{E} + \underline{G}(s)\underline{K}(s)]^{-1}$ touchs or cuts the negative real axis of the principal-gain-plane in the interval $(-\infty,-1]$.

Proof

a) Sufficient condition

This condition is

$$\det(\underline{E} + (\underline{G} + \varepsilon\Delta\underline{G})\underline{K}) \neq 0$$

or

$\lambda(\underline{E} + \underline{G}\,\underline{K} + \varepsilon\Delta\underline{G}\,\underline{K}) \neq 0$ for $0 \leq \varepsilon \leq 1$ and all s of the Nyquist-contour.

We get

$$\underline{E} + (\underline{G}\,\underline{K} + \varepsilon\Delta\underline{G}\,\underline{K}) = [\underline{E} + \varepsilon\Delta\underline{G}\,\underline{K}(\underline{E} + \underline{G}\,\underline{K})^{-1}](\underline{E} + \underline{G}\,\underline{K})$$

$(\underline{E} + \underline{G}\,\underline{K})$ has full rank and therefore the condition becomes $\lambda(\Delta\underline{G}\,\underline{K}(\underline{E} + \underline{G}\,\underline{K})^{-1}) \neq -\varepsilon^{-1}$ for $0 \leq \varepsilon \leq 1$ and all s of the Nyquist-contour.

Because this condition is fulfiled the feedback system of fig.3.21 is bibostable.

b) Necessary condition

$\underline{T}$ and $\tilde{\underline{T}}$ are supposed to be bibostable

$$\underline{T} = \underline{G}\,\underline{K}(\underline{E} + \underline{G}\,\underline{K})^{-1}$$

$$\begin{aligned}
\tilde{\underline{T}} &= (\underline{G} + \varepsilon\Delta\underline{G})\underline{K}[\underline{E} + (\underline{G} + \varepsilon\Delta\underline{G})\underline{K}]^{-1} \\
&= (\underline{G} + \varepsilon\Delta\underline{G})\underline{K}[\{\underline{E} + \varepsilon\Delta\underline{G}\,\underline{K}(\underline{E} + \underline{G}\,\underline{K})^{-1}\}(\underline{E} + \underline{G}\,\underline{K})]^{-1} \\
&= [\underbrace{\underline{G}\,\underline{K}(\underline{E} + \underline{G}\,\underline{K})^{-1}}_{\underline{T}} + \underbrace{\varepsilon\Delta\underline{G}\,\underline{K}(\underline{E} + \underline{G}\,\underline{K})^{-1}}_{\varepsilon\underline{W}}][\underline{E} + \underbrace{\varepsilon\Delta\underline{G}\,\underline{K}(\underline{E} + \underline{G}\,\underline{K})^{-1}}_{\varepsilon\underline{W}}]^{-1}
\end{aligned}$$

$$\tilde{\underline{T}} = (\underline{T} + \varepsilon\underline{W})(\underline{E} + \varepsilon\underline{W})^{-1}$$

$$\tilde{\underline{T}} = \underline{T}(\underline{E} + \varepsilon\underline{W})^{-1} + \varepsilon\underline{W}(\underline{E} + \varepsilon\underline{W})^{-1} \qquad (*)$$

The following figure shows the structure of (*)

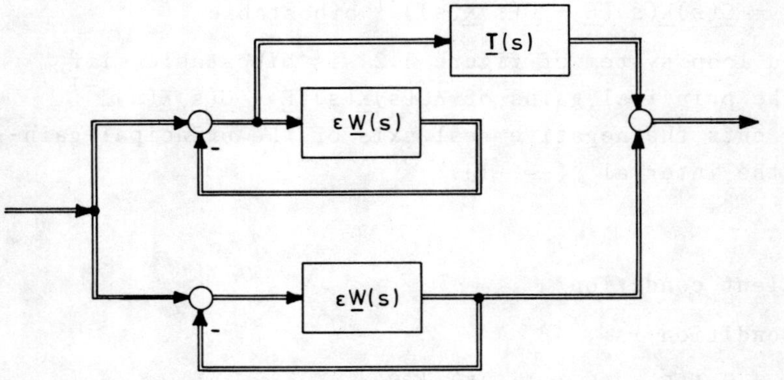

Fig.:3.22 Structure of equation (*)

This circuit is stable only, if the circuit of the next figure is stable.

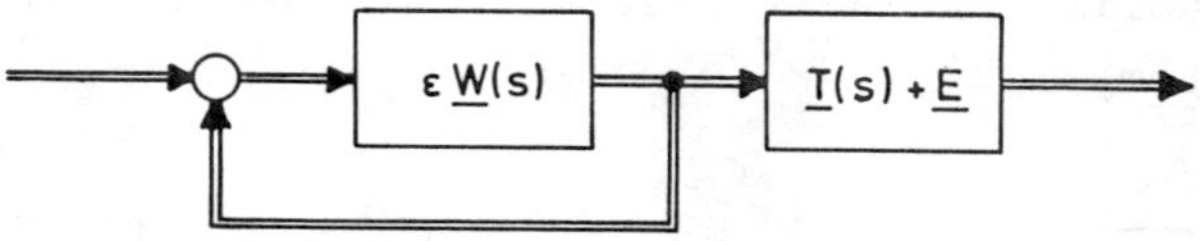

Fig.3.23: System for stability-test.

The circuit of the last figure has transfer-function

$$\tilde{\tilde{\underline{T}}} = (\underline{T}+\underline{E})\ \varepsilon\underline{W}(\underline{E} + \varepsilon\underline{W})^{-1}.$$

$\tilde{\tilde{\underline{T}}}$ is stable only, if $\varepsilon\tilde{\underline{W}} = \varepsilon\underline{W}(\underline{E}+\varepsilon\underline{W})^{-1}$ is stable for $0 \leq \varepsilon \leq 1$.
$\varepsilon\tilde{\underline{W}}$ is supposed to be stable for $0 \leq \varepsilon \leq 1$ and therefore none of the eigenvalues of $\varepsilon\underline{W}$ takes on the value -1 for any s of the Nyquist-contour and this is the same as:
All of the eigenvalues of $\Delta\underline{G}\ \underline{K}(\underline{E} + \underline{G}\ \underline{K})^{-1}$ take on values out of the intervall $(-\infty,-1]$ for all s of the Nyquist-contour.

3.3.4 MULTI-MODEL-CASE, TRANSITION STABILITY

If one tries to find the multiplicative or additiv perturbation model $\Delta\underline{G}$ for the multimodel-case normally not a bibo-stable perturbation model is found.
Therefore a further stability-theorem is given.
Suppose the closed-loop structure inclusive controller as in figure 3.24 is given.

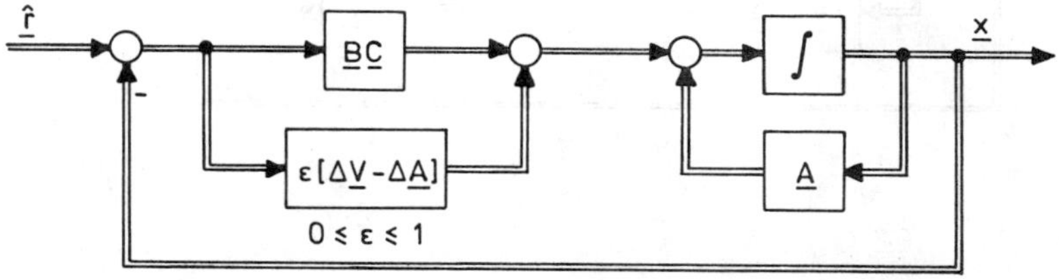

Fig.3.24: Closed-loop system.

Here $\underline{A}$, $\underline{B}$ and $\underline{C}$ describes one model of the multi-model-family and $\underline{A} + \Delta\underline{A}$, $\underline{B} + \Delta\underline{B}$ and $\underline{C} + \Delta\underline{C}$ describe another model of the multi-model-family. The next figure shows some conversions of the above system.

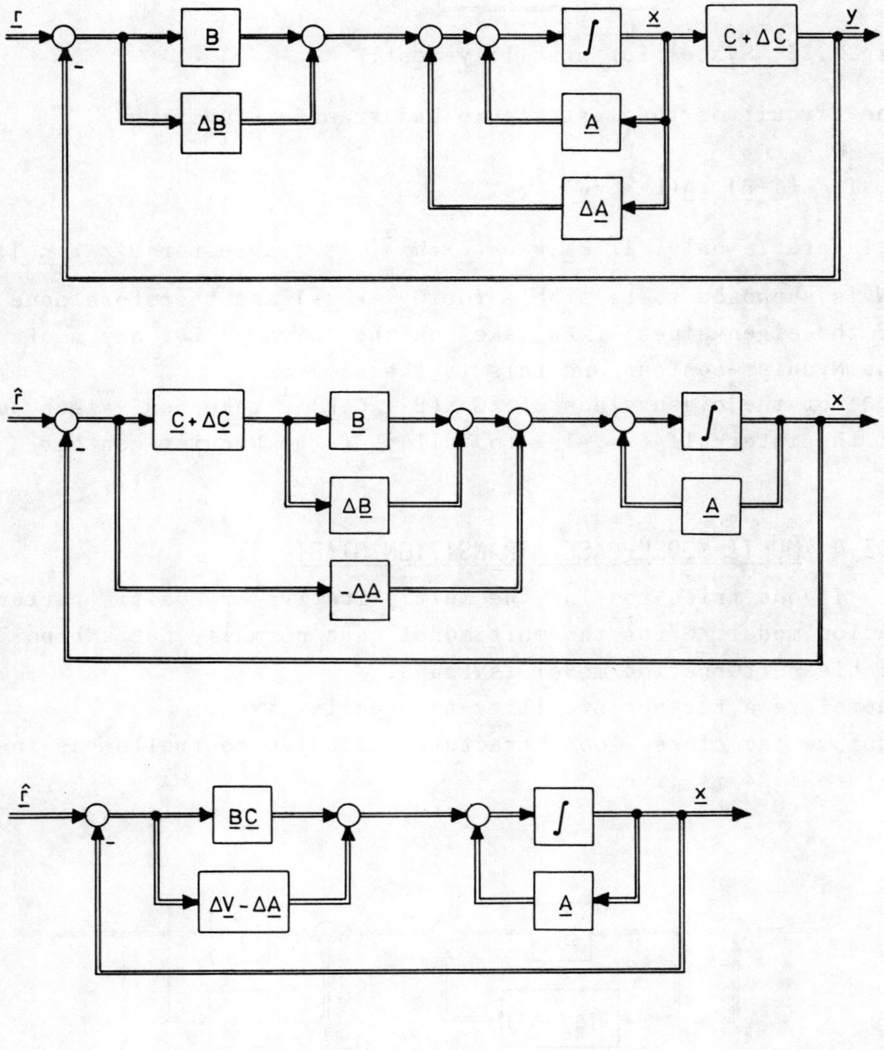

Fig.3.25: Conversion of the system of figure 3.24.

For stability-analysis the last system from figure 3.25 is separated into the nominal closed-loop system and into a feedback-path which includes only the elements of the transition, see the next figure.

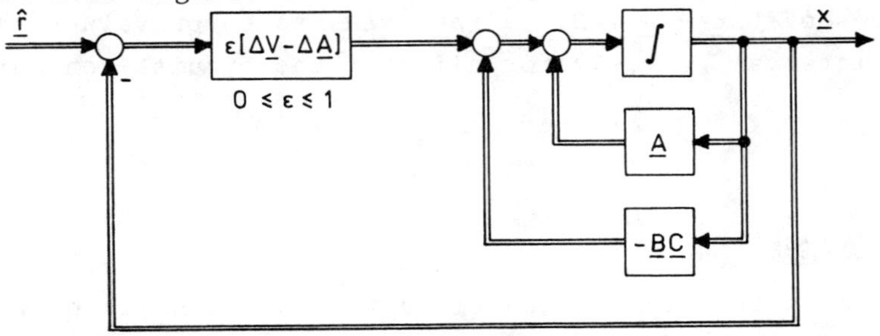

Fig.3.26: Structure for stability analysis.

The transition between the two models is given by the varying ε, $0 \leq \varepsilon \leq 1$.

THEOREM 3:

Under the condition

$$(s\underline{E} - \underline{A} + \underline{B}\ \underline{C})^{-1} \quad \text{bibostable}$$

the closed-loop-system of figure 26 is bibostable for $0 \leq \varepsilon \leq 1$ iff none of the principal gains of $(s\underline{E} - \underline{A} + \underline{B}\ \underline{C})^{-1}(\Delta\underline{V} - \Delta\underline{A})$ touchs or cuts the real axis of the principal gain-plane in the interval $(-\infty,-1]$.

Proof:

a) Sufficient condition

The stability-condition is

$$\det\{(\underline{E} + (s\underline{E} - \underline{A} + \underline{B}\ \underline{C})^{-1}[\varepsilon(\Delta\underline{V} - \Delta\underline{A})]\} \neq 0$$

for $0 \leq \varepsilon \leq 1$ and all s of the Nyquist-contour.
This is the same as all of the eigenvalues of $(s\underline{E} - \underline{A} + \underline{B}\ \underline{C})^{-1}(\Delta\underline{V} - \Delta\underline{A})$ take on values out of the interval $(-\infty,-1]$ for all s of the Nyquist-contour, i.e. the feedback system is bibostable.

b) Necessary condition

$[\underline{E} + (s\underline{E} - \underline{A} + \underline{B}\ \underline{C})^{-1} \varepsilon(\Delta\underline{V} - \Delta\underline{A})]^{-1}$

is assumed to be bibostable and therefore all of the eigenvalues of $(s\underline{E} - \underline{A} + \underline{B}\ \underline{C})^{-1}(\Delta\underline{V} - \Delta\underline{A})$ take on values out of the interval $(-\infty,-1]$ for all s of the Nyquist-contour.

3.3.5 EXAMPLE

Given is the closed-loop circuit with the helicopter from section 3.1.10. The transition from the 40 Knots-case to the 100 Knots-case is checked for transition-stability by theorem 3.

The perturbation-model is

$$\Delta\underline{V} - \Delta\underline{A} = \underline{B}_{100}\underline{C}_{100} - \underline{B}_{40}\underline{C}_{40} - \underline{A}_{100} + \underline{A}_{40} = \Delta\underline{W}$$

$$\Delta\underline{W} = \begin{bmatrix} 0.025 & -0.001 & \boxed{-10.56} & 0.072 & -1.432 & 0.049 & -0.028 & -0.353 \\ -.0896 & 0.337 & \boxed{-103.8} & -0.646 & 0.013 & -1.016 & 3.836 & 1.221 \\ 0.000 & -0.008 & 0.298 & -0.003 & -0.000 & 0.059 & -0.141 & -0.016 \\ 0.000 & 0.000 & 0.000 & 0.000 & 0.000 & 0.000 & 0.000 & -0.005 \\ 0.003 & 0.004 & -1.045 & -0.762 & 0.107 & \boxed{10.68} & 0.197 & \boxed{99.95} \\ 0.002 & 0.003 & -0.635 & -0.563 & 0.020 & -0.289 & 0.026 & -0.630 \\ 0.000 & 0.000 & -0.027 & 0.000 & 0.000 & 0.000 & 0.000 & 0.026 \\ -0.003 & 0.000 & 0.687 & 0.734 & -0.019 & -0.053 & -0.113 & 0.921 \end{bmatrix}$$

The largest parameter-variations are to be seen in the little rectangles. In figure 3.27 the principal gains of the perturbation function $(s\underline{E} - \underline{A} + \underline{B}\ \underline{C})^{-1}\Delta\underline{W}$ are shown.

None of the principal gains touchs the real axis in the critical interval. Two of the principal gains start for s = 0 with complex values. The following example shows why. Given is a transfer-matrix

$$\underline{G}(0) = \begin{bmatrix} 1 & -2 \\ 2 & 1 \end{bmatrix}.$$

The eigenvalues of this matrix are $\lambda_1 = 1 \pm j2$ and therefore the principal gains start at 1 + j2 and 1 - j2.

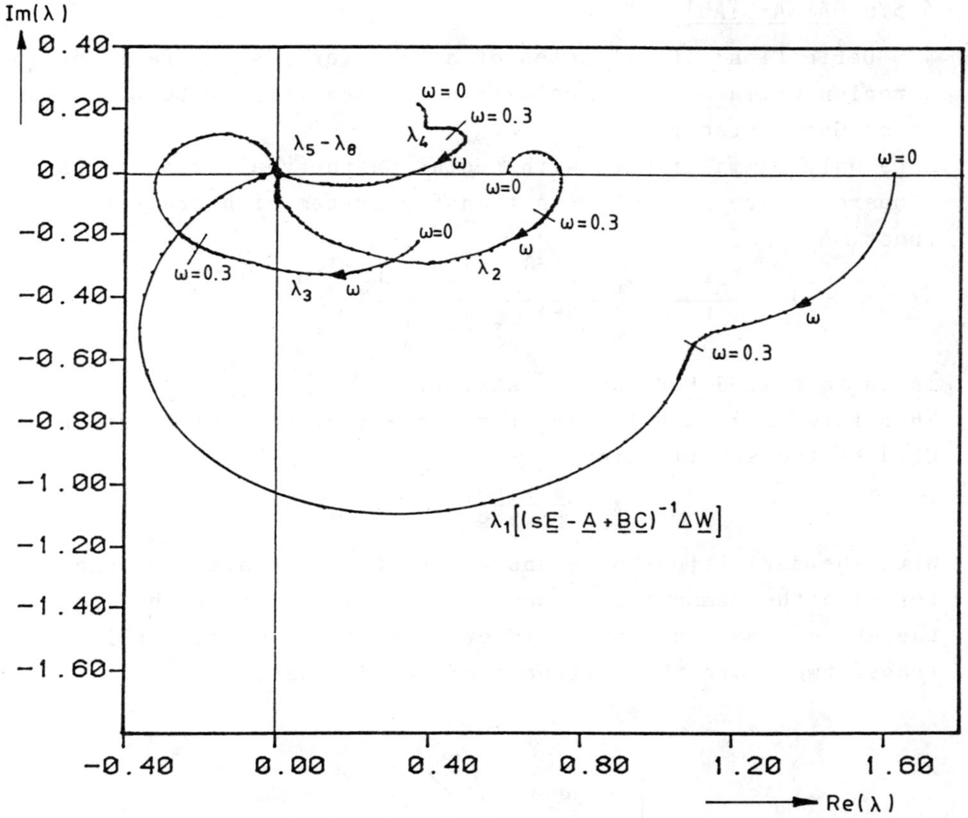

Fig.3.27: Principal gains of the helicopter-model for transition from 40 to 100 knots.

In figure 3.27 only 4 of the 8 principal gains are to be seen. This is because the values of the other 4 principal gains are so small, that they remain in the neighbourhood of the origin.

3.3.6 GAMMA-STABILITY

Definition: If all poles of a transfer-system are lying in a region Gamma of the complex-plane, then this system is said to be Gamma-stable.
Here only Gamma-circles with center on the real axis are used, compare figure 3.28. Now an transfer-system with transfer function

$$G(z) = \frac{b_n z^n + b_{n-1} z^{n-1} + \ldots + b_o}{z^n + a_{n-1} z^{n-1} + \ldots + a_o}$$

is to be tested for Gamma-stability.
Therefore first a bilinear transformation is carried out on G(z) by the substitution

$$z = \frac{a - pb}{pd - c} .$$

With the definitions of e and f out of figure 3.28 for characterizing the Gamma-region and c = 1, d = 1, a = -f, b = -e by the above substitution the inner part of the Gamma-circle is transformed into the left part of the p-plane.

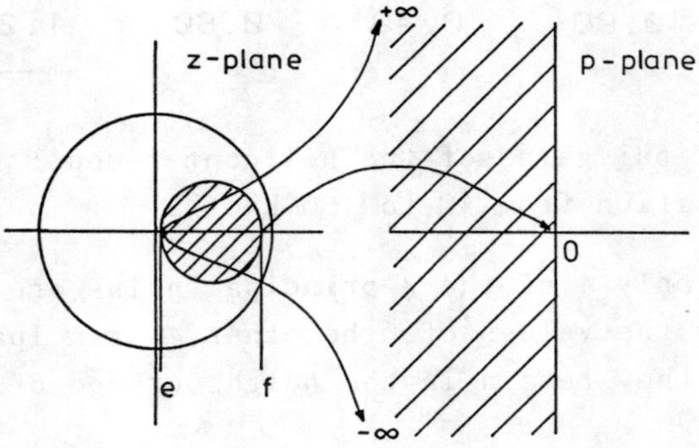

Fig.3.28: Bilinear Transformation of the Gamma-region.

Now the Gamma-stability easily can be checked with help of the Cremer-Leonhard-Michailow-or Hurwitz-Criterion.
If one has to design an Gamma-stable closed-loop control-circuit, then the Nyquist-Criterion can be used in the p-plane.

BILINEAR TRANSFORMATION [7]

Here an alternative form to the substitution method is given.
The state-space-model of the system is

$$\underline{x}(\nu+1) = \underline{\Phi}\ \underline{x}(\nu) + \underline{H}\ \underline{u}(\nu)$$

$$\underline{y}(\nu+1) = \underline{C}_z \underline{x}(\nu) + \underline{D}_z \underline{u}(\nu) .$$

The transfer-function of this system is

$$\underline{G}_z(z) = \underline{C}_z(z\underline{E} - \underline{\Phi})^{-1}\ \underline{H} + \underline{D}_z \ .$$

Now the variable z is substituted by

$$z = \frac{a - pb}{pd - c}$$

and with a few further calculations one gets for the state-space-model in the p-region

$$\dot{\underline{v}} = \underline{A}_p \underline{v} + \underline{B}_p \underline{w}$$

$$\underline{q} = \underline{C}_p \underline{v} + \underline{D}_p\ \underline{w}$$

with

$$\underline{A}_p = (b\ \underline{E} + d\ \underline{\Phi})^{-1}\ (a\ \underline{E} + c\ \underline{\Phi})$$

$$\underline{B}_p = -(b\underline{E} + d\ \underline{\Phi})^{-1}\ \underline{H}$$

$$\underline{C}_p = \underline{C}_z(d\ \underline{A}_p - c\ \underline{E})$$

$$\underline{D}_p = \underline{D}_z + d\ \underline{C}_z\ \underline{B}_p$$

and

$$\underline{G}_p(p) = \underline{C}_p(p\underline{E} + \underline{A}_p)^{-1}\underline{B}_p + \underline{D}_p .$$

This algorithm is easily transferred into an computer-programm.
There are some interesting special cases in the bilinear transformation:

$$z = \frac{a - pb}{pd - c} \quad , \quad \text{e and f see figure 3.28.}$$

a	b	c	d	
-1	1	1	1	normal bilinear transformation $z \to p$
1	1	1	-1	normal bilinear retransformation $p \to z$
-e	-f	1	1	bilinear transformation
-e	-1	f	1	bilinear retransformation
f	1	1	0	shifting of the imaginary axis. The imaginary axis after transformation now runs through the point f of the previous real axis.
0	1	k	0	"time-transformation", time runs k-times faster, in discret-time systems the radius of the "stability-region" is shifted", the new radius is k-times of the old radius
0	1	-1	0	reflection at the imaginary axis
1	0	0	1	reflection at the unit-circle.

Table 3.5: Coefficients of bilinear transformations

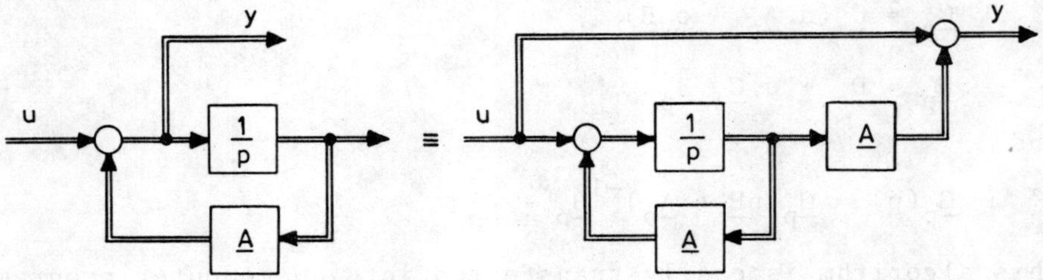

Fig.3.29: Derivation of bilinear Transformation

3.4 REFERENCES

[1] R.E.Burkard, U.Derigs
Assignement and Matching Problems
Springer-Verlag, Berlin-Heidelberg-New York,1980

[2] P.C.Müller, J.Lückel
Modale Maße für Steuerbarkeit, Beobachtbarkeit und Störbarkeit dynamischer Systeme
Zeitschrift für angewandte Mathematik und Mechanik, VOL.54,1974,S.T57-T58

[3] Model-catalogue of the VDI/VDE Arbeitskreis "Neuere theoretische Verfahren in der Regelungstechnik",1980, VDI/VDE-Gesellschaft für Meß-und Regelungstechnik (GMR), Düsseldorf

[4] R.Zurmühl
Matrizen
Springer-Verlag, Berlin-Göttingen-Heidelberg, 1964

[5] J.Doyle, G.Stein
Multivariable Feedback Design: Concepts for a Classical/Modern Synthesis
IEEE Transactions on Automatic Control AC 26 No.1, Febr.1981,pp.5-16

[6] W.Lange
Entwurf robuster Multimodell-Mehrgrößenregelkreise,
Dissertation TU Berlin, 1985

[7] B.Kouvaritakis, J.M.Edmunds
Multivariable root loci: a unified approach to finite and infinite zeros
Int.J.Control,Vol.29,No.3,pp.393-428, 1979

Index